# Aeroplanes and Dirigibles of War

## by Frederick A. Talbot

Aeroplanes and Dirigibles of War Frederick A. Talbot

PREFACE

Ever since the earliest days of the great conquest of the air, first by the dirigible balloon and then by the aeroplane, their use in time of war has been a fruitful theme for discussion. But their arrival was of too recent a date, their many utilities too unexplored to provide anything other than theories, many obviously untenable, others avowedly problematical.

Yet the part airships have played in the Greatest War has come as a surprise even to their most convinced advocates. For every expectation shattered, they have shown a more than compensating possibility of usefulness.

In this volume an endeavour has been made to record their achievements, under the stern test of trial, as an axiom of war, and to explain, in untechnical language, the many services to which they have been and may be applied.

In the preparation of the work I have received assistance from many sources--British, French, Russian and German--from official reports and from men who have played a part in the War in the Air. The information concerning German military aircraft has been obtained from Government documents, most of which were placed at my disposal before the outbreak of war.

The use of aircraft has changed the whole art and science of warfare. With its disabilities well in hand, with its strength but half revealed, the aerial service has revolutionised strategy and shorn the unexpected attack of half its terrors. The Fourth Arm is now an invaluable part of the complex military machine.

F. A. TALBOT.

CONTENTS

CHAPTER I.

The introduction of aircraft into military operations II. The military uses of

CHAPTER I

THE INTRODUCTION OF AIRCRAFT INTO MILITARY OPERATIONS

It is a curious circumstance that an invention, which is hailed as being one of the greatest achievements ever recorded in the march of civilisation, should be devoted essentially to the maiming of humanity and the destruction of property. In no other trend of human endeavour is this factor so potently demonstrated as in connection with Man's Conquest of the Air.

The dogged struggle against the blind forces of Nature was waged tenaciously and perseveringly for centuries. But the measure of success recorded from time to time was so disappointing as to convey the impression, except in a limited circle, that the problem was impossible of solution. In the meantime wondrous changes had taken place in the methods of transportation by land and sea. The steam and electric railway, steam propulsion of vessels, and mechanical movement along the highroads had been evolved and advanced to a high standard of perfection, to the untold advantage of the community. Consequently it was argued, if only a system of travel along the aerial highways could be established, then all other methods of mechanical transportation would be rendered, if not entirely obsolete, at least antiquated.

At last man triumphed over Nature--at least to such a degree as to inspire the confidence of the world at large, and to bring aerial travel and transportation within range of realisation. But what has been the result? The discovery is not devoted to the interests of peace and economic development, but to extermination and destruction.

At the same time this development may be explained. The airship and aeroplane in the present stage of evolution possess no economic value. True, cross-country cruises by airship have been inaugurated, and, up to a point, have proved popularly, if not commercially, successful, while tentative efforts have been made to utilise the aeroplane as a mail-carrier. Still, from the view-point of the community at large aerial travel is as remote as it was centuries ago.

It is somewhat interesting to observe how history is repeating itself. When the Montgolfiers succeeded in lifting themselves into the air by means of a vessel inflated with hot air, the new vehicle was hailed not so much as one possessed of commercial possibilities, but as an engine of war! When the indomitable courage and perseverance of Count von Zeppelin in the face of discouraging disasters and flagrant failures, at last commanded the attention of the German Emperor, the latter regarded the Zeppelin craft, not from the interests of peace, but as a military weapon, and the whole of the subsequent efforts of the Imperial admirer were devoted to the perfection of the airship in this one direction.

Other nations, when they embarked on an identical line of development, considered the airship from a similar point of view. In fact, outside Germany, there was very little private initiative in this field. Experiments and developments were undertaken by the military or naval, and in some instances by both branches, of the respective Powers. Consequently the aerial craft, whether it be a dirigible airship, or an aeroplane, can only be regarded from the military point of view.

Despite the achievements which have been recorded by human endeavour in the field of aerial travel, the balloon per se has by no means been superseded. It still remains an invaluable adjunct to the fighting machine. In Great Britain its value in this direction has never been ignored: of late, indeed, it has rather been developed. The captive balloon is regarded as an indispensable unit to both field and sea operations. This fact was emphasised very strongly in connection with the British naval attacks upon the German forces in Flanders, and it contributed to the discomfiture of the German hordes in a very emphatic manner.

The captive balloon may be operated from any spot where facilities exist for anchoring the paying out cable together with winding facilities for the latter. Consequently, if exigencies demand, it maybe operated from the deck of a warship so long as the latter is stationary, or even from an automobile. It is of small cubic capacity, inasmuch as it is only necessary for the bag to contain sufficient gas to lift one or two men to a height of about 500 or 600 feet.

When used in the field the balloon is generally inflated at the base, to be towed or carried forward by a squad of men while floating in the air, perhaps at a height of 10 feet. A dozen men will suffice for this duty as a rule, and in calm weather little difficulty is encountered in moving from point to point. This method possesses many advantages. The balloon can be inflated with greater ease at the base, where it is immune from interference by hostile fire. Moreover, the facilities for obtaining the requisite inflating agent--hydrogen or coal gas-- are more convenient at such a point. If the base be far removed from the spot at which it is desired to operate the balloon, the latter is inflated at a convenient point nearer the requisite position, advantage being taken of the protective covering offered by a copse or other natural obstacle.

As is well known, balloons played an important part during the siege of Paris in 1870-1, not only in connection with daring attempts to communicate with the outer world, but in reconnoitring the German positions around the beleaguered city. But this was not the first military application of the aerial vessel; it was used by the French against the Austrians in the battle of Fleurus, and also during the American Civil War. These operations, however, were of a sporadic character; they were not part and parcel of an organised military section.

It is not generally known that the British War office virtually pioneered the military use of balloons, and subsequently the methods perfected in Britain became recognised as a kind of "standard" and were adopted generally by the Powers with such modifications as local exigencies seemed to demand.

The British military balloon department was inaugurated at Chatham under Captain Templer in 1879. It was devoted essentially to the employ ment of captive balloons in war, and in 1880 a company of the Royal Engineers was detailed to the care of this work in the field. Six years previously the French military department had adopted the captive balloon under Colonel Laussedat, who was assisted among others by the well-known Captain Renard.

Germany was somewhat later in the field; the military value of captive balloons was not appreciated and taken into serious consideration here until 1884. But although British efforts were preceded by the French the latter did not develop the idea upon accepted military lines.

The British authorities were confronted with many searching problems. One of the earliest and greatest difficulties encountered was in connection with the gas for inflation. Coal gas was not always readily available, so that hydrogen had to be depended upon for the most part. But then another difficulty arose. This was the manufacture of the requisite gas. Various methods were tested, such as the electrolytic decomposition of water, the decomposition of sulphuric acid by means of iron, the reaction between slaked lime and zinc, and so forth.

But the drawbacks to every process, especially upon the field of battle, when operations have to be conducted under extreme difficulties and at high pressure, were speedily recognised. While other nations concentrated their energies upon the simplification of hydrogen-manufacturing apparatus for use upon the battle-field, Great Britain abandoned all such processes in toto. Our military organisation preferred to carry out the production of the necessary gas at a convenient manufacturing centre and to transport it, stored in steel cylinders under pressure, to the actual scene of operations. The method proved a great success, and in this way it was found possible to inflate a military balloon in the short space of 20 minutes, whereas, under the conditions of making gas upon the spot, a period of four hours or more was necessary, owing to the fact that the manufacturing process is relatively slow and intricate. The practicability of the British idea and its perfection served to establish the captive balloon as a military unit.

The British military ballooning department has always ranked as the foremost of its type among the Powers, although its work has been carried out so unostentatiously that the outside world has gleaned very little information concerning its operations. Captain Templer was an indefatigable worker and he brought the ballooning section to a high degree of efficiency from the military point of view.

But the British Government was peculiarly favoured, if such a term may be used. Our little wars in various parts of the world contributed valuable information and experience which was fully turned to account. Captive balloons for reconnoitring purposes were used by the British army for the first time at Suakim in 1885, and the section established its value very convincingly. The French military balloon department gained its first experience in this field in the previous year, a balloon detachment having been dispatched to Tonkin in 1884. In both the Tonkin and Soudan campaigns, invaluable work was accomplished by the balloon sections, with the result that this aerial vehicle has come to be regarded as an indispensable military adjunct. Indeed the activity of the German military ballooning section was directly attributable to the Anglo-French achievements therewith.

In this work, however, the British force speedily displayed its superiority and initiative. The use of compressed hydrogen was adopted, and within the course of a few years the other Powers, realising the advantages which the British department had thus obtained, decided to follow its example. The gas is stored in cylinders under a pressure varying from six to ten or more atmospheres; in other words from about 80 to 140 or more pounds per square inch. Special military wagons have been designed for the transport of these cylinders, and they are attached to the balloon train.

The balloon itself is light, and made of such materials as to reduce the weight thereof to the minimum. The British balloons are probably the smallest used by any of the Powers, but at the same time they are the most expensive. They are made of goldbeater's skin, and range in capacity from 7,000 to 10,000 cubic feet, the majority being of the former capacity. The French balloon on the other hand has a capacity exceeding 18,000 cubic feet, although a smaller vessel of 9,000 cubic feet capacity, known as an auxiliary, and carrying a single observer, is used.

The Germans, on the other hand, with their Teutonic love of the immense, favour far larger vessels. At the same time the military balloon section of the German Army eclipses that of any other nations is attached to the Intelligence Department, and is under the direct control of the General Staff. Balloon stations are dotted all over thecountry, including Heligoland and Kiel, while regular sections are attached to the Navy for operating captive balloons from warships. Although the Zeppelin and aeroplane forces have come to the front in Germany, and have relegated the captive balloon somewhat to the limbo of things that were, the latter section has never been disbanded; in fact, during the present campaign it has undergone a somewhat spirited revival.

The South African campaign emphasised the value of the British balloon section of the Army, and revealed services to which it was specially adapted, but which had previously more or less been ignored. The British Army possessed indifferent maps of the Orange Free State and the Transvaal. This lamentable deficiency was remedied in great measure by recourse to topographical photographs taken from the captive balloons. The guides thus obtained were found to be of extreme value.

During the early stages of the war the hydrogen was shipped in cylinders from the homeland, but subsequently a manufacturing plant of such capacity as to meet all requirements was established in South Africa. The cylinders were charged at this point and dispatched to the scene of action, so that it became unnecessary to transport the commodity from Britain. The captive balloon revealed the impregnability of Spion Kop, enabled Lord Roberts to ascertain the position of the Boer guns at the Battle of Paardeburg, and proved of invaluable assistance to the forces of General White during the siege of Ladysmith.

CHAPTER II

THE MILITARY USES of THE CAPTIVE BALLOON

Although the captive balloon is recognised as indispensable in military operations, its uses are somewhat limited. It can be employed only in comparatively still weather. The reason is obvious. It is essential that the balloon should assume a vertical line in relation to its winding plant upon the ground beneath, so that it may attain the maximum elevation possible: in other words, the balloon should be directly above the station below, so that if 100 yards of cable are paid out the aerostat may be 100 yards above the ground. If a wind is blowing, the helpless craft is certain to be caught thereby and driven forwards or backwards, so that it assumes an angle to its station. If this become acute the vessel will be tilted, rendering the position of the observers somewhat precarious, and at the same time observing efficiency will be impaired.

This point may be appreciated more easily by reference to the accompanying diagram. A represents the ground station and B the position of the captive balloon when sent aloft in calm weather, 300 feet of cable being paid out. A wind arises and blows the vessel forward to the position C. At this point the height of the craft in relation to the ground has been reduced, and the reduction must increase proportionately as the strength of the wind increases and forces the balloon still more towards the ground. At the same time, owing to the tilt given to the car, observation is rendered more difficult and eventually becomes extremely dangerous.

A wind, if of appreciable strength, develops another and graver danger. Greater strain will be imposed upon the cable, while if the wind be gusty, there is the risk that the vessel will be torn away from its anchoring rope and possibly lost. Thus it will be seen that the effective utilisation of a captive balloon is completely governed by meteorological conditions, and often it is impossible to use it in weather which exercises but little influence upon dirigibles or aeroplanes.

The captive balloon equipment comprises the balloon, together with the observer's basket, the wire-cable whereby it is anchored and controlled, and the winding apparatus. Formerly a steam engine was necessary for the paying in and out of the cable, but nowadays this is accomplished by means of a petrol-driven motor, an oil-engine, or even by the engine of an automobile. The length of cable varies according to the capacity of the balloon and the maximum operating height.

The average British balloon is able to lift about 290 or 300 pounds, which may be taken to represent the weight of two observers. On the other hand, the French and German balloons are able to carry four times this weight, with the exception of the French auxiliaries, which are designed to lift one observer only. The balloons of the two latter Powers have also a greater maximum altitude; it is possible to ascend to a height of some 2,000 feet in one of these.

The observing station is connected with the winding crew below either by a telephone, or some other signalling system, the method practised varying according to circumstances. In turn the winding station is connected with the officer in charge of the artillery, the fire of which the captive balloon is directing. The balloon observer is generally equipped with various instruments, such as telescope, photographic cameras, and so forth, so as to be able, if necessary, to prepare a topographical survey of the country below. By this means the absence of reliable maps may be remedied, or if not regarded, as sufficiently correct they may be checked and counter-checked by the data gained aloft.

Seeing that the gas has to be transported in cylinders, which are weighty, it is incumbent that the waste of this commodity should be reduced to the minimum. The balloon cannot be deflated at night and re-inflated in the morning--it must be maintained in the inflated condition the whole time it is required for operation.

There are various methods of consummating this end. One method is to haul in the balloon and to peg it down on all sides, completing the anchorage by the attachment of bags filled with earth to the network. While this process is satisfactory in calm weather, it is impracticable in heavy winds, which are likely to spring up suddenly. Consequently a second method is practised. This is to dig a pit into the ground of sufficient size to receive the balloon. When the latter is hauled in it is lowered into this pit and there pegged down and anchored. Thus it is perfectly safe during the roughest weather, as none of its bulk is exposed above the ground level. Furthermore it is not a conspicuous object for the concentration of hostile fire.

In some instances, and where the military department is possessed of an elaborate equipment such as characterises the German army, when reconnaissance is completed and the balloon is to be removed to another point, the gas is pumped back into the cylinders for further use. Such an economical proceeding is pretty and well adapted to manoeuvres, but it is scarcely feasible in actual warfare, for the simple reason that the pumping takes time. Consequently the general procedure, when the balloon has completed its work, is to permit the gas to escape into the air in the usual manner, and to draw a fresh supply of gas from further cylinders when the occasion arises for re-inflation.

Although the familiar spherical balloon has proved perfectly adequate for reconnoitring in the British and French armies, the German authorities maintained that it was not satisfactory in anything but calm weather. Accordingly scientific initiative was stimulated with a view to the evolution of a superior vessel. These endeavours culminated in the Parseval-Siegsfeld captive balloon, which has a quaint appearance. It has the form of a bulky cylinder with hemispherical extremities. At one end of the balloon there is a surrounding outer bag, reminiscent of a cancerous growth. The lower end of this is open. This attachment serves the purpose of a ballonet. The wind blowing against the opening, which faces it, charges the ballonet with air. This action, it is claimed, serves to steady the main vessel, somewhat in the manner of the tail of a kite, thereby enabling observations to be made as easily and correctly in rough as in calm weather. The appearance of the balloon while aloft is certainly curious. It appears to be rearing up on end, as if the extremity saddled with the ballonet were weighted.

British and French captive balloon authorities are disposed to discount the steadying effect of this attachment, and, indeed, to maintain that it is a distinct disadvantage. It may hold the vessel steadier for the purpose of observation, but at the same time it renders the balloon a steadier target for hostile fire. On the other hand, the swaying of a spherical balloon with the wind materially contributes to its safety. A moving object, particularly when its oscillations are irregular and incalculable, is an extremely difficult object at which to take effective aim.

Seeing that even a small captive balloon is of appreciable dimensions--from 25 to 33 feet or more in diameter--one might consider it an easy object to hit. But experience has proved otherwise. In the first place the colour of the balloon is distinctly protective. The golden or yellowish tinge harmonises well with the daylight, even in gloomy weather, while at night-time it blends excellently with the moonlight. For effective observations a high altitude is undesirable. At a height of 600 feet the horizon is about 28 miles from the observer, as compared with the 3 miles constituting the range of vision from the ground over perfectly flat country. Thus it will be seen that the "spotter" up aloft has the command of a considerable tract.

Various ways and means of finding the range of a captive balloon have been prepared, and tables innumerable are available for committal to memory, while those weapons especially designed for aerial targets are fitted with excellent range-finders and other instruments. The Germans, with characteristic thoroughness, have devoted considerable attention to this subject, but from the results which they have achieved up to the present this guiding knowledge appears to be more spectacular and impressive than effective.

To put a captive balloon out of action one must either riddle the envelope, causing it to leak like a sieve, blow the vessel to pieces, or ignite the highly inflammable gas with which it is inflated. Individual rifle fire will inflict no tangible damage. A bullet, if it finds its billet, will merely pass through the envelope and leave two small punctures. True, these vents will allow the gas to escape, but this action will proceed so slowly as to permit the vessel to remain aloft long enough to enable the observer to complete his work. A lucky rifle volley, or the stream of bullets from a machine gun may riddle the envelope, precipitating a hurried descent, owing to the greater number of perforations through which the gas is able to escape, but as a rule the observer will be able to land safely.

Consequently the general practice is to shatter the aerostat, and to this end either shrapnel, high explosive, or incendiary shells will be used. The former must explode quite close to the balloon in order to achieve the desired end, while the incendiary shell must actually strike it, so as to fire the gas. The high explosive shell may explode effectually some feet away from the vessel, inasmuch as in this instance dependence is placed upon the terrific concussion produced by the explosion which, acting upon the fragile fabric of the balloon, brings about a complete collapse of the envelope. If a shrapnel is well placed and explodes immediately above the balloon, the envelope will be torn to shreds and a violent explosion of the gas will be precipitated. But as a matter of fact, it is extremely difficult to place a shrapnel shell so as to consummate this end. The range is not picked up easily, while the timing of the fuse to bring about the explosion of the shell at the critical moment is invariably a complex problem.

One favourite method of finding the range of a balloon is shown in the accompanying diagrams. The artillery battery is at B and the captive balloon, C, is anchored at A. On either side of B and at a specified distance, observers O1 and O2 respectively are stationed. First a shell is fired at "long" range, possibly the maximum range of the gun. It bursts at D. As it has burst immediately in the line of sight of B, but with the smoke obscured by the figure of the balloon C, it is obvious to B that the explosion has occurred behind the objective, but at what distance he cannot tell. To O1 and

O2,however, it is seen to have burst at a considerable distance behind C though to the former it appears to have burst to the left and to the second observer to the right of the target.

Another shell, at "short" range, is now fired, and it bursts at E. The explosion takes place in the line of sight of B, who knows that he has fired short of the balloon because the latter is eclipsed by the smoke. But the two observers see that it is very short, and here again the explosion appears to O1 to have occurred to the right of the target, while to O2 it has evidently burst to the left of the aerostat, as revealed by the relation of the position of the balloon to the bursting of the shell shown in Fig. 3.

A third round is fired, and the shell explodes at F. In this instance the explosion takes place below the balloon. Both the observers and the artillery man concur in their deductions upon the point at which the shell burst. But the shell must explode above the balloon, and accordingly a fourth round is discharged and the shell bursts at G.

This appears to be above the balloon, inasmuch as the lines of sight of the two observers and B converge at this point. But whether the explosion occurs immediately above the vessel as is desired, it is impossible to say definitely, because it may explode too far behind to be effective. Consequently, if this shell should prove abortive, the practice is to decrease the range gradually with each succeeding round until the explosion occurs at the critical point, when, of course, the balloon is destroyed. An interesting idea of the difficulty of picking up the range of a captive balloon may be gathered from the fact that some ten minutes are required to complete the operation.

But success is due more to luck than judgment. In the foregoing explanation it is premised that the aerial vessel remains stationary, which is an ex tremely unlikely contingency. While those upon the ground are striving to pick up the range, the observer is equally active in his efforts to baffle his opponents. The observer follows each successive, round with keen interest, and when the shells appear to be bursting at uncomfortably close quarters naturally he intimates to his colleagues below that he desires his position to be changed, either by ascending to a higher point or descending. In fact, he may be content to come to the ground. Nor must the fact be overlooked that while the enemy is trying to place the observer hors de combat, he is revealing the position of his artillery, and the observer is equally industrious in picking up the range of the hostile guns for the benefit of his friends below.

When the captive balloon is aloft in a wind the chances of the enemy picking up the range thereof are extremely slender, as it is continually swinging to and fro. While there is always the possibility of a shell bursting at such a lucky moment as to demolish the aerial target, it is generally conceded to be impossible to induce a shell to burst within 100 yards of a balloon, no matter how skilfully the hostile battery may be operated.

The value of the captive balloon has been demonstrated very strikingly throughout the attack upon the entrenched German positions in Flanders. Owing to the undulating character of the dunes the "spotters" upon the British monitors and battle ships are unable to obtain a sweeping view of the country. Accordingly captive balloons are sent aloft in some cases from the deck of the monitors, and in others from a suitable point upon the beach itself. The aerial observer from his point of vantage is able to pick up the positions of the German forces and artillery with ease and to communicate the data thus gained to the British vessels, although subjected to heavy and continuous hostile fire. The difficulty of hitting a captive balloon has been graphically emphasised, inasmuch as the German artillerists have failed to bring down a solitary balloon. On the other hand the observer in the air is able to signal the results of each salvo fired from the British battleships as they manoeuvre at full speed up and down the coastline, while he keeps the fire of the monitors concentrated upon the German positions until the latter have been rendered untenable or demolished. The accuracy of the British gun-fire has astonished even the Germans, but it has been directly attributable to the rangefinder perched in the car of the captive balloon and his rapid transmission of information to the vessels below.

The enthusiastic supporters of aerial navigation maintained that the dirigible and the aeroplane would supersede the captive balloon completely. But as a matter of fact the present conflict has established the value of this factor more firmly than ever. There is not the slightest possibility that the captive balloon sections of the belligerents will be disbanded, especially those which have the fruits of experience to guide them. The airship and the aeroplane have accomplished wonders, but despite their achievements the captive balloon has fully substantiated its value as a military unit in its particular field of operations.

CHAPTER III

GERMANY'S RISE TO MILITARY AIRSHIP SUPREMACY

Two incidents in the history of aviation stand out with exceptional prominence. The one is the evolution of the Zeppelin airship--a story teeming with romance and affording striking and illuminating glimpses of dogged perseverance, grim determination in the face of repeated disasters, and the blind courageous faith of the inventor in the creation of his own brain. The second is the remarkable growth of Germany's military airship organisation, which has been so rapid and complete as to enable her to assume supremacy in this field, and that within the short span of a single decade.

The Zeppelin has always aroused the world's attention, although this interest has fluctuated. Regarded at first as a wonderful achievement of genius, afterwards as a freak, then as the ready butt for universal ridicule, and finally with awe, if not with absolute terror--such in brief is the history of this craft of the air.

Count von Zeppelin can scarcely be regarded as an ordinary man. He took up the subject of flight at an age which the majority of individuals regard as the opportune moment for retirement from activity, and, knowing nothing about mechanical engineering, he concentrated his energies upon the study of this science to enable him to master the difficulties of a mechanical character incidental to the realisation of his grand idea. His energy and indomitable perseverance are equalled by his ardent patriotism, because, although the Fatherland discounted his idea when other Powers were ready to consider it, and indeed made him tempting offers for the acquisition of his handiwork, he stoutly declined all such solicitations, declaring that his invention, if such it may be termed, was for his own country and none other.

Count von Zeppelin developed his line of study and thought for one reason only. As an old campaigner and a student of military affairs he realised the shortcomings of the existing methods of scouting and reconnoitring. He appreciated more than any other man of the day perhaps, that if the commander-in-chief of an army were provided with facilities for gazing down upon the scene of operations, and were able to take advantage of all the information accruing to the man above who sees all, he would hold a superior position, and be able to dispose his forces and to arrange his plan of campaign to the most decisive advantage. In other words, Zeppelin conceived and developed his airship for one field of application and that alone-military operations. Although it has achieved certain successes in other directions these have been subsidiary to the primary intention, and have merely served to emphasise its military value.

Von Zeppelin was handicapped in his line of thought and investigation from the very first. He dreamed big things upon a big scale. The colossal always makes a peculiar and irresistible appeal to the Teutonic nature. So he contemplated the perfection of a big dirigible, eclipsing in every respect anything ever attempted or likely to be attempted by rival countries. Unfortunately, the realisation of the "colossal" entails an equally colossal financial reserve, and the creator of this form of airship for years suffered from financial cramp in its worst manifestation. Probably it was to the benefit of the world at large that Fortune played him such sorry tricks. It retarded the growth of German ambitions in one direction very effectively.

As is well known Zeppelin evolved what may be termed an individual line of thought in connection with his airship activities. He adopted what is known as the indeformable airship: that is to say the rigid, as opposed to the semi-rigid and flexible craft. As a result of patient experiment and continued researches he came to the conclusion that a huge outer envelope taking the form of a polygonal cylinder with hemispherical ends, constructed upon substantial lines with a metallic skeleton encased within an impermeable skin, and charged with a number of smaller balloon-shaped vessels containing the lifting agent--hydrogen gas--would fulfil his requirements to the greatest advantage. Model after model was built upon these lines. Each was subjected to searching tests with the invariable result attending such work with models. Some fulfilled the expectations of the inventor, others resolutely declined to illustrate his reasonings in any direction.

The inevitable happened. When a promising model was completed finally the inventor learned to his sorrow what every inventor realises in time. His fortune and the resources of others had been poured down the sink of experiment. To carry the idea from the model to the practical stage required more money, and it was not forthcoming. The inventor sought to enlist the practical sympathy of his country, only to learn that in Germany, as in other lands, the axiom concerning the prophet, honour, and country prevails. No exuberant inventor received such a cold douche from a Government as did Count Zeppelin from the Prussian authorities. For two years further work was brought practically to a standstill: nothing could be done unless the sinews of war were forthcoming. His friends, who had assisted him financially with his models, now concluded that their aid had been misplaced.

The inventor, though disappointed, was by no means cast down. He clung tenaciously to his pet scheme and to such effect that in 1896 a German Engineering Society advanced him some funds to continue his researches. This support sufficed to keep things going for another two years, during which time a full-sized vessel was built. The grand idea began to crystallise rapidly, with the result that when a public company was formed in 1898, sufficient funds were rendered available to enable the first craft to be constructed. It aroused considerable attention, as well it might, seeing that it eclipsed anything which had previously been attempted in connection with dirigibles. It was no less than 420 feet in length, by 38 feet in diameter, and was fitted with two cars, each of which carried a sixteen horse-power motor driving independent propellers rigidly attached to the body of the vessel. The propellers were both vertical and horizontal, for the purpose of driving the ship in the two planes--vertical and horizontal respectively.

The vessel was of great scientific interest, owing to the ingenuity of its design and construction. The metallic skeleton was built up from aluminium and over this was stretched the fabric of the envelope, care being observed to reduce skin friction, as well as to achieve impermeability. But it was the internal arrangement of the gas-lifting balloons which provoked the greatest concern. The hull was divided into compartments, each complete in itself, and each containing a small balloon inflated with hydrogen. It was sub-division as practised in connection with vessels ploughing the water applied to aerial craft, the purpose being somewhat the same. As a ship of the seas will keep afloat so long as a certain number of its subdivisions remain watertight, so would the Zeppelin keep aloft if a certain number of the gas compartments retained their charges of hydrogen. There were no fewer than seventeen of these gas-balloons arranged in a single line within the envelope. Beneath the hull and extending the full length of the latter was a passage which not only served as a corridor for communication between the cars, but also to receive a weight attached to a cable worked by a winch. By the movement of this weight the bow or stem of the vessel could be tilted to assist ascent and descent.

The construction of the vessel subsequently proved to be the easiest and most straightforward part of the whole undertaking. There were other and more serious problems to be solved. How would such a monster craft come to earth? How could she be manipulated upon the ground? How could she be docked? Upon these three points previous experience was silent. One German inventor who likewise had dreamed big things, and had carried them into execution, paid for his temerity and ambitions with his life, while his craft was reduced to a mass of twisted and torn metal. Under these circumstances Count Zeppelin decided to carry out his flights over the waters of the Bodensee and to house his craft within a floating dock. In this manner two uncertain factors might be effectively subjugated.

Another problem had been ingeniously overcome. The outer envelope presented an immense surface to the atmosphere, while temperature was certain to play an uncertain part in the behaviour of the craft. The question was to reduce to the minimum the radiation of heat and cold to the bags containing the gas. This end was achieved by leaving a slight air space between the inflated gas balloons and the inner surface of the hull.

The first ascent was made on July 2nd, 1900, but was disappointing, several breakdowns of the mechanism occurring while the vessel was in mid-air, which rendered it unmanageable, although a short flight was made which sufficed to show that an independent speed of 13 feet per second could be attained. The vessel descended and was made fast in her dock, the descent being effected safely, while manoeuvring into dock was successful. At least three points about which the inventor had been in doubt appeared to be solved--his airship could be driven through the air and could be steered; it could be brought to earth safely; and it could be docked.

The repairs to the mechanism were carried out and on October 17th and 21st of the same year further flights were made. By this time certain influential Teuton aeronautical experts who had previously ridiculed Zeppelin's idea had made a perfect volte-face. They became staunch admirers of the system, while other meteorological savants participated in the trials for the express purpose of ascertaining just what the ship could do. As a result of elaborate trigonometrical calculations it was ascertained that the airship attained an independent speed of 30 feet per second, which exceeded anything previously achieved. The craft proved to be perfectly manageable in the air, and answered her helm, thus complying with the terms of dirigibility. The creator was flushed with his triumph, but at the same time was doomed to experience misfortune. In its descent the airship came to "earth" with such a shock that it was extensively damaged. The cost of repairing the vessel was so heavy that the company declined to shoulder the liability, and as the Count was unable to defray the expense the wreck was abandoned.

Although a certain meed of success had been achieved the outlook seemed very black for the inventor. No one had any faith in his idea. He made imploring appeals for further money, embarked upon lecturing campaigns, wrote aviation articles for the Press, and canvassed possible supporters in the effort to raise funds for his next enterprise. Two years passed, but the fruits of the propaganda were meagre. It was at this juncture, when everything appeared to be impossible, that Count Zeppelin discovered his greatest friend. The German Emperor, with an eye ever fixed upon new developments, had followed Zeppelin's uphill struggle, and at last, in 1902, came to his aid by writing a letter which ran:--

"Since your varied flights have been reported to me it is a great pleasure to me to express my acknowledgment of your patience and your labours, and the endurance with which you have pressed on through manifold hindrances till success was near. The advantages of your system have given your ship the greatest attainable speed and dirigibility, and the important results you have obtained have produced an epoch-making step forward in the construction of airships and leave laid down a valuable basis for future experiments."

This Imperial appreciation of what had been accomplished proved to be the turning point in the inventor's fortunes. It stimulated financial support, and the second airship was taken in hand. But misfortune still pursued him. Accidents were of almost daily occurrence. Defects were revealed here and weaknesses somewhere else. So soon as one trouble was overcome another made itself manifest. The result was that the whole of the money collected by his hard work was expended before the ship could take to the air. A further crash and blasting of cherished hopes appeared imminent, but at this moment another Royal personage came to the inventor's aid.

The King of Wurtemberg took a personal interest in his subject's uphill struggle, and the Wurtemberg Government granted him the proceeds of a lottery. With this money, and with what he succeeded in raising by hook and by crook, and by mortgaging his remaining property, a round L20,000 was obtained. With this capital a third ship was taken in hand, and in 1905 it was launched. It was a distinct improvement upon its predecessors. The airship was 414 feet in length by 38 feet in diameter, was equipped with 17 gas balloons having an aggregate capacity of 367,000 cubic feet of hydrogen, was equipped with two 85 horse-power motors driving four propellers, and displaced 9 tons. All the imperfections incidental to the previous craft had been eliminated, while the ship followed improved lines in its mechanical and structural details.

The trials with this vessel commenced on November 30th, 1905, but ill-luck had not been eluded. The airship was moored upon a raft which was to be towed out into the lake to enable the dirigible to ascend. But something went wrong with the arrangements. A strong wind caught the ungainly airship, she dipped her nose into the water, and as the motor was set going she was driven deeper into the lake, the vessel only being saved by hurried deflation.

Six weeks were occupied in repairs, but another ascent was made on January 17th, 1906. The trials were fairly satisfactory, but inconclusive. One of the motors went wrong, and the longitudinal stability was found to be indifferent. The vessel was brought down, and was to be anchored, but the Fates ruled otherwise. A strong wind caught her during the night and she was speedily reduced to indistinguishable scrap.

Despite catastrophe the inventor wrestled gamely with his project. The lessons taught by one disaster were taken to heart, and arrangements to prevent the recurrence thereof incorporated in the succeeding craft. Unfortunately, however, as soon as one defect was remedied another asserted itself. It was this persistent revelation of the unexpected which caused another period of indifference towards his invention. Probably nothing more would have been heard of the Zeppelin after this last accident had it not been for the intervention of the Prussian Government at the direct instigation of the Kaiser, who had now taken Count Zeppelin under his wing. A State lottery was inaugurated, the proceeds of which were handed over to the indefatigable inventor, together with an assurance that if he could keep aloft 24 hours without coming to earth in the meantime, and could cover 450 miles within this period, the Government would repay the whole of the money he had lavished upon his idea, and liquidate all the debts he had incurred in connection therewith.

Another craft was built, larger than its predecessors, and equipped with two motors developing 170 horse-power. Upon completion it was submitted to several preliminary flights, which were so eminently successful that the inventor decided to make a trial trip under conditions closely analogous to those imposed for the Government test. On June 20th, 1908, at 8:26 a.m. the craft ascended and remained aloft for 12 hours, during which time it made an encouraging circular tour. Flushed with this success, the Count considered that the official award was within reach, and that all his previous disasters and misfortunes were on the eve of redemption.

The crucial test was essayed on August 5th, 1908. Accompanied by twelve observers the vessel ascended and travelled without incident for eight hours. Then a slight mishap demanded attention, but was speedily repaired, and was ignored officially as being too trivial to influence the main issue. Victory appeared within measurable distance: the arduous toil of many patient years was about to be rewarded. The airship was within sight of home when it had to descend owing to the development of another motor fault. But as it approached the ground, Nature, as if infuriated at the conquest, rose up in rebellion. A sudden squall struck the unwieldy monster. Within a few moments it became unmanageable, and through some inscrutable cause, it caught fire, with the result that within a few moments it was reduced to a tangled mass of metallic framework.

It was a catastrophe that would have completely vanquished many an inventor, but the Count was saved the gall of defeat. His flight, which was remarkable, inasmuch as he had covered 380 miles within 24 hours, including two unavoidable descents, struck the Teuton imagination. The seeds so carefully planted by the "Most High of Prussia" now bore fruit. The German nation sympathised with the indomitable inventor, appreciated his genius, and promptly poured forth a stream of subscriptions to enable him to build another vessel. The intimation that other Powers had approached the Count for the acquisition of his idea became known far and wide, together with the circumstance that he had unequivocally refused all offers. He was striving for the Fatherland, and his unselfish patriotism appealed to one and all. Such an attitude deserved hearty national appreciation, and the members of the great German public emptied their pockets to such a degree that within a few weeks a sum of L300,000 or $1,500,000 was voluntarily subscribed.

All financial embarrassments and distresses were now completely removed from the Count's mind. He could forge ahead untrammelled by anxiety and worry. Another Zeppelin was built and it created a world's record. It remained aloft for 38 hours, during which time it covered 690 miles, and, although it came to grief upon alighting, by colliding with a tree, the final incident passed unnoticed. Germany was in advance of the world. It had an airship which could go anywhere, irrespective of climatic conditions, and in true Teuton perspective the craft was viewed from the military standpoint. Here was a means of obtaining the mastery of the air: a formidable engine of invasion and aerial attack had been perfected. Consequently the Grand Idea must be supported with unbounded enthusiasm. The Count was hailed by his august master as "The greatest German of the twentieth century," and in this appreciation the populace wholeheartedly concurred. Whether such a panegyric from such an auspicious quarter is praise indeed or the equivalent of complete condemnation, history alone will be able to judge, but when one reflects, at this moment, upon the achievements of this aircraft during the present conflagration, the unprejudiced will be rather inclined to hazard the opinion that Imperial Teuton praise is a synonym for damnation.

Although the Zeppelin was accepted as a perfect machine it has never been possible to disperse the atmosphere of disaster with which it has been enveloped from the first. Vessel after vessel has gone up in smoke and flame: few craft of this type have enjoyed more than an evanescent existence; and each successive catastrophe has proved more terrible than its predecessor. But the Teutonic nation has been induced to pin its whole faith on this airship, notwithstanding that the more levelheaded engineers of other countries have always maintained the craft to be a "mechanical monstrosity" condemned from its design and principles of construction to disaster. Unshaken by this adverse criticism, Germany rests assured that by means of its Zeppelins it will achieve that universal supremacy which it is convinced is its Destiny.

This blind child-like faith has been responsible for the establishment and development of the Zeppelin factories. At Friedrichshafen the facilities are adequate to produce two of these vessels per month, while another factory of a similar capacity has been established at Berlin. Unfortunately such big craft demand large docks to accommodate them, and in turn a large structure of this character constitutes an easy mark for hostile attack, as the raiding airmen of the Allies have proved very convincingly.

But the Zeppelin must not be under-rated. Magnificent performances have been recorded by these vessels, such as the round 1,000 miles' trip in 1909, and several other equally brilliant feats since that date. It is quite true that each astounding achievement has been attended by an equally stupendous accident, but that is accepted as a mere incidental detail by the faithful Teutonic nation. Many vivid prophecies of the forthcoming flights by Zeppelin have been uttered, and it is quite probable that more than one will be fulfilled, but success will be attributable rather to accident than design.

Although the Zeppelin is the main stake of the German people in matters pertaining to aerial conquest, other types of airships have not been ignored, as related in another chapter. They have been fostered upon a smaller but equally effective scale. The semi-rigid Parseval and Gross craft have met with whole-hearted support, since they have established their value as vessels of the air, which is tantamount to the acceptance of their military value.

The Parseval is pronounced by experts to be the finest expression of aeronautical engineering so far as Teuton effort is concerned. Certainly it has placed many notable flights to its credit. The Gross airship is an equally serviceable craft, its lines of design and construction closely following those of the early French supple airships. There are several other craft which have become more or less recognised by the German nation as substantial units of war, such as the Ruthemberg, Siemens-Schukert, and so forth, all of which have proved their serviceability more or less conclusively. But in the somewhat constricted Teuton mind the Zeppelin and the Zeppelin only represents the ultima Thule of aerial navigation and the means for asserting the universal character of Pan-Germanism as well as "Kultur."

CHAPTER IV

AIRSHIPS OF WAR

So much has been said and written concerning the Zeppelin airship, particularly in its military aspect, that all other developments in this field have sunk into insignificance so far as the general public is concerned. The Zeppelin dirigible has come to be generally regarded as the one and only form of practical lighter-than-air type of aircraft. Moreover, the name has been driven home with such effect that it is regarded as the generic term for all German airships.

These are grievous fallacies. The Zeppelin is merely one of a variety of types, even in Germany, although at the moment it probably ranks as the solitary survivor of the rigid system of construction. At one time, owing to the earnestness with which the advantages of this form of design were discussed, and in view of the fact that the Zeppelin certainly appeared to triumph when all other designs failed, Great Britain was tempted to embrace the rigid form of construction. The building of an immense vessel of this class was actively supported and it was aptly christened the "May-fly." Opponents of the movement tempered their emphatic condemnatory criticism so far as to remark that it MAY FLY, but as events proved it never did. The colossal craft broke its back before it ever ventured into the air, and this solitary experience proving so disastrous, the rigid form of construction was abandoned once and for all. The venture was not in vain; it brought home to the British authorities more convincingly than anything else that the Zeppelin was a mechanical monstrosity. The French never even contemplated the construction of such a craft at that time, estimating it at its true value, and the British failure certainly served to support French antagonism to the idea. Subsequently, however, an attempt at rigid construction was made in France with the "Spiess" airship, mainly as a concession to public clamour.

Even in Germany itself the defects of the Zeppelin were recognised and a decided effort to eliminate them was made by Professor Schutte in co-operation with a manufacturer of Mannheim named Lanz. The joint product of their ambitions, the Schutte-Lanz, is declared to be superior to the Zeppelin, but so far it has failed to justify any of the claims of its designers. This vessel, which also favours the colossal, is likewise of the rigid type, but realising the inherent dangers accruing from the employment of metal for the framework, its constructors have used wood, reinforced and strengthened where necessary by metallic angle-iron, plates, and bracing; this utilisation of metal is, however, carried out very sparingly. The first vessel of this class was a huge failure, while subsequent craft have not proved much more successful.

In fact, one of the largest German airships ever designed, L4, is, or rather was, a Schutte-Lanz, with a capacity of 918,000 cubic feet, but over 6,000 pounds lighter than a Zeppelin of almost similar dimensions. I say "was" since L4 is no more. The pride of its creators evinced a stronger preference for Davy Jones' Locker than its designed realm. Yet several craft of this type have been built and have been mistaken for Zeppelins owing to the similarity of the broad principles of design and their huge dimensions. In one vital respect they are decidedly inferior to their contemporary--they are not so speedy.

The most successful of the German lighter-than-air machines are those known respectively as the semi rigid and non-rigid types, the best examples of which are the Gross and Parseval craft. Virtually they are Teutonic editions of the successful French craft of identical design by which they were anticipated. The Lebaudy is possibly the most famous of the French efforts in this direction. The gas-bag has an asymmetrical shape, and is pointed at both ends, although the prow is blunter or rounder than the stem. The gas-bag comprises a single chamber for the inflating agent, the distended shape of the envelope being sustained by means of an air-ballonet. By varying the contents of the latter through the agency of a pump the tension of the gas in the lifting envelope can be maintained, and the shape of the inflated balloon preserved under all conditions.

Beneath the gas-bag is a long strengthened girder, and from this in turn the car is suspended. It is the introduction of this rigid girder which is responsible for the descriptive generic term of "semi-rigid." On the other hand the "non-rigid" type may be roughly described as a pisciform balloon fitted with propelling machinery, inasmuch as the car containing the driving machinery is suspended from the balloon in the manner of the car in the ordinary drifting vessel. So far as the French effort is concerned the Bayard-Clement type is the best example of the non-rigid system; it is represented in Germany by the Parseval class.

The Gross airship has been definitely adopted as a military machine by the

German authorities, and figures in the "M" class. The "M-IV" completed in 1913 is the largest of this type, and differs from its prototypes in that it carries two cars, each fitted with motors, whereas the earlier machines were equipped with a single gondola after the French pattern. This vessel measures 320 feet in length, has a maximum diameter of 44 1/2 feet, displaces 13 tons, and is fitted with motors developing 450 horse-power, which is sufficient to give it a speed of 47 miles per hour. This vessel represents a huge advance upon its predecessors of this design, inasmuch as thelatter were about 245 feet in length by 36 1/4 feet in diameter, and displaced only six tons, while the single car was provided with a motor developing only 150 horse-power, the speed being 28 miles per hour. Thus it will be seen that a huge development has suddenly taken place, a result due no doubt to the co-operation of the well-known engineer Basenach. The "M-IV" is essentially an experiment and great secrecy has been maintained in regard to the trials which have been carried out therewith, the authorities merely vouchsafing the fact that the airship has proved completely successful in every respect; conclusive testimony of this is offered by the inclusion of the vessel in the active aerial fleet of Germany.

But it is the Parseval which is regarded as the finest type of airship flying the German flag. This vessel is the product of slow evolution, for it is admitted to be a power-driven balloon. Even the broad lines of the latter are preserved, the shape being that of a cylinder with rounded ends. It is the direct outcome of the "Drachen-Balloon," perfected by Parseval and Siegsfeld, the captive balloon which is an indispensable part of the German military equipment.

The complete success of the suspension system in this captive balloon prompted Parseval to continue his researches and experiments in regard to the application of power to the vessel, so as to induce it to move independently of the wind. The suspension system and the car are the outstanding features of the craft. It is non-rigid in the strictest interpretation of the term, although, owing to the incorporation of the steadying hollow "mattress" (as it is called by its inventor), the strength of the suspension system, and the substantial character of the car, it conveys an impression of great solidity. The thinnest rope, both manilla and steel, in the suspension system is as thick as a man's finger, while the car, measuring 30 feet in length by 6 feet in width, carried out in wood, is a striking example of the maximum of strength with the minimum of weight, being as steady and as solid as a boat's deck. The propellers are collapsible, although in the latest craft of this class they are semi-rigid.

The mechanical equipment is also interesting. There are two propellers, and two motors, each nominally driving one propeller. But should one motor break down, or motives of economy, such as husbanding of fuel, render it advisable to run upon one engine, then the two propellers may be driven by either of the motors.

The inventor has perfected an ingenious, simple, and highly efficient coupling device to attain this end, but to ensure that the propeller output is of the maximum efficiency in relation to the engine, the pitch of the propellers may be altered and even reversed while the engine is running. When one motor only is being used, the pitch is lowered until the propellers revolve at the speed which they would attain if both engines were in operation. This adjustment of the propeller pitch to the most economical engine revolutions is a distinctive characteristic, and contributes to the efficiency and reliability of the Parseval dirigible to a very pronounced degree.

Steering in the vertical plane is also carried out upon distinctive lines. There are no planes for vertical steering, but movement is accomplished by tilting the craft and thus driving the gas from one end of the balloon to the other. This is effected by the manipulation of the air-ballonets, one of which is placed at the prow and stem of the gas bag respectively. If it is desired to descend the gas is driven from the forward to the after end of the envelope, merely by inflating the bow ballonet with air by means of a pump placed in the car. If ascent is required, the after-ballonet is inflated, thereby driving the gas to the forward end of the balloon, the buoyancy of which is thus increased. The outstanding feature of the "Drachen-Balloon" is incorporated in the airship. This is the automatic operation of the safety valve on the gas-bag directly by the air ballonets. If these ballonets empty owing to the pressure of the gas within the envelope, a rope system disposed within the balloon and connecting the ballonets and the gas-valve at the top is stretched taut, thereby opening the gas-valve. In this manner the gas-pressure becomes reduced until the ballonets are enabled to exercise their intended function. This is a safety precaution of inestimable value.

The Parseval is probably the easiest dirigible to handle, inasmuch as it involves no more skill or knowledge than that required for an ordinary free balloon. Its movements in the vertical plane are not dissimilar to those of the aeroplane, inasmuch as ascent and descent are normally conducted in a "screwing" manner, the only exception being of course in abrupt descent caused by the ripping of the emergency-valve. On one occasion, it is stated, one of the latest machines of this type, when conducting experimental flights, absolutely refused to descend, producing infinite amusement both among the crowd and those on board.

The development of the Parseval is directly attributable to the influence and intimate interest of the Kaiser, and undoubtedly this represents the wisest step he ever made in the realm of aeronautics. It certainly has enabled the German military machine to become possessed of a significant fleet of what may be described as a really efficient and reliable type of dirigible. The exact number of military Parsevals in commission is unknown, but there are several classes thereof, in the nature of aerial cruisers and vedettes.

The largest and most powerful class are those known as the B type, measuring about 240 feet in length by 40 feet maximum diameter, of 223,000 cubic feet capacity, and fitted with two motorsand two propellers. This vessel carries about 10 passengers, can climb to a maximum height of approximately 8,500 feet, and is capable of remaining in the air for twenty hours upon a single fuel charge. While this is the largest and most serviceable type of Parseval designed for military duties, there is another, the A class, 200 feet in length with accommodation for six passengers in addition to the crew of three, which is capable of attaining a maximum altitude of 6,700 feet, and has an endurance capacity of 15 hours. This class also is fitted with twin propellers and motors. In addition there are the C and E classes, carrying from four to eight passengers, while the vedettes are represented by the D and F classes, which have a maximum altitude of 2,000 feet and can remain aloft for only five hours upon a single fuel charge. These smaller vessels, however, have the advantage of requiring only one or two men to handle them. The present military Parseval dirigible is made in one of these five standardised classes, experience having established their efficiency for the specified military services for which they are built. In point of speed they compare favourably with the latest types of Zeppelin, the speeds of the larger types ranging from 32 to 48 miles per hour with a motor effort of 360 to 400 horse-power.

So far as the French airships of war are concerned, the fleet is somewhat heterogeneous, although the non-rigid type prevails. The French aerial navy is represented by the Bayard-Clement, Astra, Zodiac, and the Government-built machines. Although the rigid type never has met with favour in France, there is yet a solitary example of this system of construction--the Spiess, which is 460 feet in length by 47 feet in diameter and has a displacement of 20 tons. The semi-rigid craft are represented by the Lebaudy type, the largest of which measures 293 feet in length by 51 feet in diameter, and has a displacement of 10 tons.

One may feel disposed to wonder why the French should be apparently backward in this form of aerial craft, but this may be explained by the fact that the era of experiment had not been concluded at the time war was declared, with the result that it has been somewhat difficult to determine which type would meet the military requirements of the country to the best advantage. Moreover, the French military authorities evinced a certain disposition to relegate the dirigible to a minor position, convinced that it had been superseded by the heavierthan-air machine. Taken on the whole, the French airship fleet is inferior to the German in point of speed, if not numerically, but this deficiency is more than counterbalanced by the skill and ability of the men manning their craft, who certainly are superior to their contemporaries in Germany, combined with the proved character of such craft as are in service.

The same criticism may be said to apply to Great Britain. That country was backward in matters pertaining to the airship, because its experiments were carried out spasmodically while dependence was reposed somewhat too much upon foreign effort. The British airships are small and of low speed comparatively speaking. Here again it was the advance of the aeroplane which was responsible for the manifestation of a somewhat indifferent if not lethargic feeling towards the airship. Undoubtedly the experiments carried out in Great Britain were somewhat disappointing. The one and only attempt to out-Zeppelin the Zeppelin resulted in disaster to the craft before she took to the air, while the smaller craft carried out upon far less ambitious lines were not inspiritingly successful. Latterly the non-rigid system has been embraced exclusively, the craft being virtually mechanically driven balloons. They have proved efficient and reliable so far as they go, but it is the personal element in this instance also which has contributed so materially to any successes achieved with them.

But although Great Britain and France apparently lagged behind the Germans, appreciable enterprise was manifested in another direction. The airship was not absolutely abandoned: vigilance was maintained for a superior type of craft. It was an instance of weighing the advantages against the disadvantages of the existing types and then evolving for a design which should possess the former without any of the latter. This end appears to be achieved with the Astra type of dirigible, the story of the development of which offers an interesting chapter in the annals of aeronautics.

In all lighter-than-air machines the resistance to the air offered by the suspension ropes is considerable, and the reduction of this resistance has proved one of the most perplexing problems in the evolution of the dirigible. The air is broken up in such a manner by the ropes that it is converted into a brake or drag with the inevitable result that the speed undergoes a severe diminution. A full-rigged airship such as the Parseval, for instance, may present a picturesque appearance, but it is severely unscientific, inasmuch as if it were possible to eliminateor to reduce the air-resistance offered by the ropes, the speed efficiency might be raised by some sixty per cent and that without any augmentation of the propelling effort. As a matter of fact Zeppelin solved this vexatious problem unconsciously. In his monster craft the resistance to the air is reduced to a remarkable degree, which explains why these vessels, despite all their other defects are able to show such a turn of speed.

It was this feature of the Zeppelin which induced Great Britain to build the May-fly and which likewise induced the French Government to stimulate dirigible design and construction among native manufacturers, at the same time, however, insisting that such craft should be equal at least in speed to the Zeppelins. The response to this invitation was the Spiess, which with its speed of 45 miles per hour ranked, until 1914, as one of the fastest dirigibles in the French service.

In the meantime a Spanish engineer, Senor Torres, had been quietly working out a new idea. He realised the shortcomings of the prevailing types of airships some eleven years ago, and unostentatiously and painstakingly set out to eliminate them by the perfection of a new type of craft. He perfected his idea, which was certainly novel, and then sought the assistance of the Spanish Government. But his fatherland was not adapted to the prosecution of the project. He strove to induce the authorities to permit even a small vessel to be built, but in vain. He then approached the French Astra Company. His ambition was to build a vessel as large as the current Zeppelin, merely to emphasise the value of his improvement upon a sufficiently large scale, and to enable comparative data concerning the two designs to be obtained. But the bogey of expense at first proved insuperable. However, the French company, decided to give the invention a trial, and to this end a small "vedette" of about 53,000 cubic feet displacement was built.

Although an unpretentious little vessel, it certainly served to emphasise the importance of the Torres idea. It was pitted against the "Colonel Renard," the finest ship at that time in the French aerial service, which had proved the fastest airship in commission, and which also was a product of the Astra Company. But this fine craft was completely outclassed by the puny Astra-Torres.

The builders and the inventor were now additionally anxious to illustrate more emphatically the features of this design and to build a far larger vessel. The opportunity was offered by the British Government, which had been following the experiments with the small Astra-Torres in France. An order was given for a vessel of 282,500 cubic feet displacement; in this instance it was ranged against another formidable rival--the Parseval. But the latter also failed to hold its own against the Spanish invention, inasmuch as the Astra-Torres built for the British authorities exceeded a speed of 50 miles per hour in the official tests. This vessel is still doing valuable duty, being attached to the British air-service in France.

The achievements of the British vessel were not lost upon the French Government, which forthwith placed an order for a huge vessel of 812,200 cubic feet capacity, equipped with motors developing 1,000 horse-power, which it was confidently expected would enable a speed of 60 miles per hour to be attained. Thus France would be able to meet the Germans upon fairly level terms, inasmuch as the speed of the latest Zeppelins does not exceed 60 miles per hour. So confident were the authorities that a second order for an even larger vessel was placed before the first large craft was completed.

This latter vessel is larger than any Zeppelin yet built, seeing that it displaces 38 tons, and is fitted with motors developing 1,000 horse-power. It has recently been completed, and although the results of the trials, as well as the dimensions of the craft have not been published, it is well known that the speed has exceeded 60 miles per hour, so that France now possesses the speediest dirigible in the world.

The Torres invention has been described as wonderful, scientifically perfect and extremely simple. The vessel belongs to the non-rigid class, but the whole of the suspension system is placed within the gas-bag, so that the air-resistance offered by ropes is virtually eliminated in its entirety, for the simple reason that practically no ropes are placed outside the envelope. The general principle of design may be gathered from the accompanying diagram. It is as if three sausage-shaped balloons were disposed pyramidally--two lying side by side with one super-imposed, with the bags connected at the points where the circular sections come into contact. Thus the external appearance of the envelope is decidedly unusual, comprising three symmetrical ridges. At the points where the three bags come into contact cloth bands are stretched across the arcs, thereby forming a cord. The suspension system is attached to the upper corners of the inverted triangle thus formed, and converges in straight lines through the gas space. The bracing terminates in collecting rings from which a short vertical cable extends downwards through a special accordion sleeve to pass through the lower wall of the envelope. These sleeves are of special design, the idea being to permit the gas to escape under pressure arising from expansion and at the same time to provide ample play for the cable which is necessary in a flexible airship.

This cable emerges from the envelope only at the point or points where the car or cars is or are placed. In the British airship of this type there is only one car, but the larger French vessels are equipped with two cars placed tandem-wise. The vertical cable, after extending downwards a certain distance, is divided, one rope being attached to one, and the second to the other side of the car. The two-bladed propellers are disposed on either side of the car, in each of which a 500 horse-power motor is placed.

The Astra-Torres type of dirigible may be said to represent the latest expression in airship design and construction. The invention has given complete satisfaction, and has proved strikingly successful. The French Government has completed arrangements for the acquisition of larger and more powerful vessels of this design, being now in the position to contest every step that is made by Germany in this field. The type has also been embraced by the Russian military authorities. The Astra-Torres airship has a rakish appearance, and although the lines of the gas-bag are admitted to increase frictional resistance, this is regarded as a minor defect, especially when the many advantages of the invention are taken into consideration.

CHAPTER V

GERMANY'S AERIAL DREADNOUGHT FLEET

Although Germany, as compared with France, was relatively slow to recognise the immense possibilities of aircraft, particularly dirigibles, in the military sense, once the Zeppelin had received the well-wishes of the Emperor William, Teuton activities were so pronounced as to enable the leeway to be made up within a very short while. While the Zeppelin commanded the greatest attention owing to the interesting co-operation of the German Emperor, the other types met with official and royal recognition and encouragement as already mentioned. France, which had held premier position in regard to the aerial fleet of dirigibles for so long, was completely out-classed, not only in dimensions but also in speed, as well as radius of action and strategical distribution of the aerial forces.

The German nation forged ahead at a great pace and was able to establish a distinct supremacy, at least on paper. In the light of recent events it is apparent that the German military authorities realised that the dawn of "The Day" was approaching rapidly, and that it behoved them to be as fully prepared in the air as upon the land. It was immaterial that the Zeppelin was the synonym for disaster. By standardisation its cost could be reduced while construction could be expedited. Furthermore, when the matter was regarded in its broadest aspect, the fact was appreciated that forty Zeppelins could be built at the cost of one super-Dreadnought, so that adequate allowance could be made for accidents now and then, since a Zeppelin catastrophe, no matter how complete it may be, is regarded by the Teuton as a mere incident inseparable from progressive development.

At the beginning of the year 1914 France relied upon being strengthened by a round dozen new dirigibles. Seven of these were to be of 20,000 cubic metres' capacity and possessed of a speed of 47 miles per hour. While the existing fleet was numerically strong, this strength was more apparent than real, for the simple reason that a large number of craft were in dry-dock undergoing repair or overhaul while many of the units were merely under test and could not be regarded therefore as in the effective fleet. True, there were a certain number of private craft which were liable to be commandeered when the occasion arose, but they could not be considered as decided acquisitions for the simple reason that many were purely experimental units.

Aerial vessels, like their consorts upon the water, have been divided into distinctive classes. Thus there are the aerial cruisers comprising vessels exceeding 282,000 cubic feet in capacity; scouts which include those varying between 176,600 and 282,000 cubic feet capacity; and vedettes, which take in all the small or mosquito craft. At the end of 1913, France possessed only four of the first-named craft in actual commission and thus immediately available for war, these being the Adjutant Vincenot, Adjutant Reau, Dupuy de Lome, and the Transaerien. The first three are of 197,800 cubic feet. All, however, were privately owned.

On the other hand, Germany had no fewer than ten huge vessels, ranging from 353,000 to 776,900 cubic feet capacity, three of which, the Victoria Luise, Suchard, and Hansa, though owned privately, were immediately available for war. Of these the largest was the Zeppelin naval vessel "L-1" 525 feet in length, by 50 feet diameter, of 776,900 cubic feet capacity, equipped with engines developing 510 horse-power, and with a speed of 51.8 miles per hour.

At the end of 1913 the effective aerial fleet of Germany comprised twenty large craft, so far in advance of the French aerial cruisers as to be worthy of the name bestowed upon them-- "Aerial Dreadnoughts." This merely represented the fleet available for immediate use and did not include the four gigantic Suchard-Schutte craft, each of 847,500 cubic feet, which were under construction, and which were being hurried forward to come into commission early in 1914.

But the most interesting factor, apart from the possession of such a huge fleet of dirigible air-craft, was their distribution at strategical points throughout the Empire as if in readiness for the coming combat. They were literally dotted about the country. Adequate harbouring facilities had been provided at Konigsberg, Berlin, Posen, Breslau, Kiel, Hamburg, Wilhelmshaven, Dusseldorf, Cologne, Frankfort, Metz, Mannheim, Strasburg, and other places, with elaborate headquarters, of course, at Friedrichshafen upon Lake Constance. The Zeppelin workshops, harbouring facilities, and testing grounds at the latter point had undergone complete remodelling, while tools of the latest type had been provided to facilitate the rapid construction and overhaul of the monster Zeppelin dirigibles. Nothing had been left to chance; not an item was perfunctorily completed. The whole organisation was perfect, both in equipment and operation. Each of the above stations possessed provision for an aerial Dreadnought as well as one or more aerial cruisers, in addition to scouts or vedettes.

Upon the outbreak of hostilities Germany's dirigible fleet was in a condition of complete preparedness, was better organised, and better equipped than that of any of her rivals. At the same time it constituted more of a paper than a fighting array for reasons which I will explain later. But there was another point which had escaped general observation. Standardisation of parts and the installation of the desired machinery had accomplished one greatly desired end--the construction of new craft had been accelerated. Before the war an interesting experiment was carried out to determine how speedily a vessel could be built. The result proved that a dirigible of the most powerful type could be completed within eight weeks and forthwith the various constructional establishments were brought into line so as to maintain this rate of building.

The growth of the Zeppelin, although built upon disaster, has been amazing. The craft of 1906 had a capacity of 430,000 cubic feet and a speed of 36 miles per hour. In 1911 the creator of this type launched a huge craft having a capacity of 627,000 cubic feet. In the meantime speed had likewise been augmented by the use of more powerful motors until 52 miles an hour was attained. But this by no means represented the limit. The foregoing vessels had been designed for land service purely and simply, but now the German authorities demanded similar craft for naval use, possessed of high speed and greater radius of action. Count Zeppelin rose to the occasion, and on October

7th, 1912, launched at Friedrichshafen the monster craft "L-I," 525 feet in length, 50 feet in diameter, of 776,900 cubic feet capacity, a displacement of 22 tons and equipped with three sets of motors aggregating more than 500 horse-power, and capable of imparting a speed of 52 miles per hour.

The appearance of this craft was hailed with intense delight by the German nation, while the naval department considered her to be a wonderful acquisition, especially after the searching reliability trial. In charge of Count Zeppelin and manned by a crew of 22 officers and men together with nearly three tons of fuel--the fuel capacity conveys some idea of her possible radius of action--she travelled from Friedrichshafen to Johannisthal in 32 hours. On this remarkable journey another point was established which was of far-reaching significance. The vessel was equipped with wireless telegraphy and therewith she kept in touch with the earth below throughout the journey, dropping and picking up wireless stations as she progressed with complete facility. This was a distinct achievement, inasmuch as the vessel having been constructed especially for naval operations she would be able to keep in touch with the warships below, guiding them unerringly during their movement.

The cross-country trip having proved so completely successful the authorities were induced to believe that travelling over water would be equally satisfactory. Accordingly the "L-I" was dispatched to the island of Heligoland, the intention being to participate in naval manoeuvres in order to provide some reliable data as to the value of these craft operating in conjunction with warships. But in these tests German ambition and pride received a check. The huge Zeppelin was manoeuvring over the North Sea within easy reach of Heligoland, when she was caught by one of those sudden storms peculiar to that stretch of salt water. In a moment she was stricken helpless; her motive power was overwhelmed by the blind forces of Nature. The wind caught her as it would a soap-bubble and hurled her into the sea, precipitating the most disastrous calamity in the annals of aeronautics, since not only was the ship lost, but fifteen of her crew of 22 officers and men were drowned.

The catastrophe created consternation in German aeronautical circles. A searching inquiry was held to explain the disaster, but as usual it failed to yield much material information. It is a curious circumstance, but every successive Zeppelin disaster, and their number is legion, has been attributable to a new cause. In this instance the accident was additionally disturbing, inasmuch as the ship had been flying across country continuously for about twelve months and had covered more miles than any preceding craft of her type. No scientific explanation for the disaster was forthcoming, but the commander of the vessel, who sank with his ship, had previously ventured his personal opinion that the vessel was over-loaded to meet the calls of ambition, was by no means seaworthy, and that sooner or later she would be caught by a heavy broadside wind and rendered helpless, or that she would make a headlong dive to destruction. It is a significant fact that he never had any faith in the airship, at least for sea duty, though in response to official command he carried out his duties faithfully and with a blind resignation to Fate.

Meantime, owing to the success of the "L-I" in cross-country operations, another and more powerful craft, the "L-II" had been taken in hand, and this was constructed also for naval use. While shorter than her consort, being only 487 feet over all, thisvessel had a greater beam--55 feet. This latter increase was decided because it was conceded to be an easier matter to provide for greater beam than enhanced length in the existing air-ship harbours. The "L-II" displaced 27 tons--five tons in excess of her predecessor. In this vessel many innovations were introduced, such as the provision of the passage-way connecting the cars within the hull, instead of outside the latter as had hitherto been the practice, while the three cars were placed more closely together than formerly. The motors were of an improved type, giving an aggregate output of 900 horse-power, and were divided into four separate units, housed in two engine-rooms, the front car being a replica in every detail of the navigating bridge of a warship.

This vessel was regarded as a distinct improvement upon the "L-I," although the latter could boast some great achievements. But her glory was short-lived. In the course of the Government trials, while some 900 feet aloft, the huge vessel suddenly exploded and was burned in the air, a mass of broken and twisted metal-work falling to the ground. Of the 28 officers and men, including members of the Admiralty Board who were conducting the official trials, all but one were killed outright, and the solitary exception was so terribly burned as to survive the fall for only a few hours.

The accident was remarkable and demonstrated very convincingly that although Count Zeppelin apparently had made huge strides in aerial navigation through the passage of years, yet in reality he had made no progress at all. He committed the identical error that characterised the effort of Severo Pax ten years previously, and the disaster was directly attributable to the self-same cause as that which overwhelmed the Severo airship. The gas, escaping from the balloons housed in the hull, collected in the confined passage-way communicating with the cars, came into contact with a naked light, possibly the exhaust from the motors, and instantly detonated with terrific force, blowing the airship to fragments and setting fire to all the inflammable materials.

In this airship Zeppelin committed an unpardonable blunder. He had ignored the factor of "internal safety," and had deliberately flown in the face of the official rule which had been laid down in France after the Severo disaster, which absolutely forbade the inclusion of such confined spaces as Zeppelin had incorporated. This catastrophe coming so closely as it did upon the preceding disaster to the pride of the German aerial fleet somewhat shook public confidence in these craft, while aeronautical authorities of other countries described the Zeppelin more vehemently than ever as a "mechanical monstrosity" and a "scientific curiosity."

The Zeppelin has come to be feared in a general manner, but this result is due rather to stories sedulously circulated, and which may be easily traced to Teutonic sources. Very few data of a reliable character have been allowed to filter through official circles. We have been told somewhat verbosely of what it can accomplish and of its high degree of efficiency and speed. But can credence be placed in these statements?

When Zeppelin IV made its unexpected descent at Luneville, and was promptly seized by the French authorities, the German War office evinced distinct signs of uneasiness. The reason was speedily forth coming. The captain of the craft which had been captured forgot to destroy his log and other records of data concerning the vessel which had been scientifically collected during the journey. All this information fell into the hands of the French military department, and it proved a wondrous revelation. It enabled the French to value the Zeppelin at its true worth, which was by no means comparable to the estimate based on reports skilfully circulated for the benefit of the world at large.

Recently the French military department permitted the results of their expert official examination to be made public. From close investigation of the log-book and the diagrams which had been prepared, it was found that the maximum speed attained by Zeppelin IV during this momentous flight was only 45 miles per hour! It was ascertained, moreover, that the load was 10,560 pounds, and the ascensional effort 45,100 pounds. The fuel consumption had averaged 297 pounds per hour, while the fuel tanks carried sufficient for a flight of about seven hours. The airship had attained a maximum height of about 6,230 feet, to reach which 6,600 pounds of ballast had to be discarded. Moreover, it was proved that a Zeppelin, if travelling under military conditions with full armament and ammunition aboard, could carry sufficient fuel for only ten hours at the utmost, during which, if the slightest head-wind prevailed, it could not cover more than 340 miles on the one fuel charge.

This information has certainly proved a revelation and has contributed to the indifference with which the Parisians regard a Zeppelin raid. At the outbreak of war the Zeppelin station nearest to Paris was at Metz, but to make the raid from that point the airship was forced to cover a round 500 miles. It is scarcely to be supposed that perfectly calm weather would prevail during the whole period of the flight, so that a raid would be attended by considerable risk. That this handicap was recognised in German military circles is borne out by the fact that a temporary Zeppelin hangar was established at a point considerably nearer the French capital, for the purpose of enabling a raid to be carried out with a greater possibility of success.

The capture of Zeppelin IV revealed another important fact. The critical flying height of the airship is between 3,300 and 4,000 feet. To attempt a raid at such an altitude would be to court certain disaster, inasmuch as the vessel would have to run the gauntlet of the whole of the French artillery, which it is admitted has a maximum range exceeding the flying altitude of the Zeppelin. That the above calculation is within reason is supported by the statements of Count Zeppelin himself, who has declared that his airships are useless at a height exceeding 5,000 feet. Confirmatory evidence upon this point is offered by the raid upon the British East Coast towns, when it is stated that the aircraft were manoeuvring at a height not exceeding 2,000 feet.

CHAPTER VI

THE MILITARY VALUE OF GERMANY'S AERIAL FLEET

Although the Zeppelin undoubtedly has been over-rated by the forces to which it is attached, at the same time it must not be under-estimated by its detractors. Larger and more powerful vessels of this type have been, and still are being, constructed, culminating, so far as is known, in the "L-5," which is stated to have a capacity of about 1,000,000 cubic feet, and to possess an average speed of 65 miles per hour.

While it is generally maintained that the Zeppelins will prove formidable in attack, greater reliance is being placed upon the demoralising or terrifying effect which they are able to exercise. Owing to the fact that from 3 to 5 tons of fuel--say 900 to 1,500 gallons of gasoline or petrol--can be carried aboard, giving them a wide radius of action, it is doubtful whether they could travel from Cologne to London and back upon a single fuel charge, since such a raid would entail a journey of about 600 miles. The latest types of this craft are said to possess a high ascensional speed, which offers a distinct protection against aeroplane attack. According to such official information as has been vouchsafed, a Zeppelin, when hard pressed, is able to rise vertically 3,500 feet in about three minutes. This is far in excess of the ascensional speed of even the speediest aeroplane. of course, the penalty for such a factor has to be paid: the loss of gas is appreciable and may lead to the craft's ultimate undoing. At the same time, however, it is able to maintain the superior position as compared with the aeroplane for a considerable period: the upper reaches of the air are its sanctuary.

Nor must the nocturnal activities of the Zeppelin be overlooked. So far as night operations by these vessels are concerned, little has leaked out, so that the possibilities of the airship in this direction are still somewhat hypothetical. The fact remains, however, that it is night movements which perhaps are the most to be dreaded by the enemy. According to official German sources of information the latest types of Zeppelins are engined by "noiseless" motors. There is nothing remarkable in this feature, since the modern motor-car virtually answers to this description, although in this instance quietness is obtained for the most part by recourse to the sleeve-valve engine. Still, the ordinary Otto-cycle internal combustion engine can be rendered almost silent by the utilisation of adequate muffling devices, which, in the Zeppelin, are more possible of incorporation than in the aeroplane, because the extra weight imposed by this acquisition is a minor consideration in comparison with the lifting power of the vessel.

Night operations, however, have not proved eminently successful. The very darkness which protects the aerial prowler also serves a similar purpose in connection with its prey. But aerial operations under the cover of darkness are guided not so much by the glare of lights from below as betrayal by sound. The difference between villages and cities may be distinguished from aloft, say at 1,500 to 3,000 feet, by the hum which life and movement emit, and this is the best guide to the aerial scout or battleship. The German authorities have made a special study of this peculiar problem, and have conducted innumerable tests upon the darkest nights, when even the sheen of the moon has been unavailable, for the express purpose of training the aerial navigators to discover their position from the different sounds reaching them from below. In other words, the corsair in the skies depends more upon compass and sound than upon compass and vision when operating after dark. The searchlights with which the Zeppelins are equipped are provided merely for illuminating a supposed position. They are not brought into service until the navigator concludes that he has arrived above the desired point: the ray of light which is then projected is merely to assist the crew in the discharge of the missiles of destruction.

The Zeppelin, however, owing to its speed, both in the horizontal and vertical planes, is essentially a unit for daylight operations. The other airships which Germany possesses, and which for the most part are of the non-rigid type, are condemned to daylight operations from the character of their design. Owing to their low speeds they may be dismissed as impossible aerial vessels for hazardous work and are not regarded by the German authorities as all-round airships of war.

Craft of the air are judged in Germany from the one standard only. This may be a Teutonic failing, but it is quite in keeping with the Teutonic spirit of militarism. Commercialism is a secondary factor. To the German Emperor an airship is much what a new manufacturing process or machine is to the American. Whereas the latter asks, "How much will it save me on the dollar?"

to the War Lord of Germany--and an airship notwithstanding its other recommendatory features is judged solely from this standpoint--the question is "What are its military qualifications?"

When the semi-rigid airship "V-I" was brought before the notice of the German military department the pressing point concerning its military recommendations arose at once. The inventor had foreseen this issue and was optimistic. Thereupon the authorities asked if the inventor were prepared to justify his claims. The retort was positive. Forthwith the Junkers decided to submit it to the test.

This ship is of quite a distinctive type. It is an aerial cruiser, and the inventor claims that it combines all the essential qualifications of the Zeppelin and of the competitors of the latter, in addition to the advantage of being capable of dissection, transportation in parts, and rapid re-erection at any desired spot. The length of the vessel is about 270 feet; maximum diameter approximately 42 feet, and capacity about 300,000 cubic feet. The outstanding feature is a rigid keel-frame forming a covered passage way below the envelope or gas-bag, combined with easy access to all parts of the craft while under way, together with an artificial stiffening which dispenses with the necessity of attaching any additional cars. The frame is so designed that the load, as well as the ballast and fuel tanks, may be distributed as desired, and at the same time it ensures an advantageous disposition of the steering mechanism, far removed from the centre of rotation at the stern, without any overloading of the latter.

The lifting part of the airship comprises a single gas bag fitted with two ballonets provided to ensure the requisite gas-tension in the main envelope, while at the same time permitting, in times of emergency, a rapid change of altitude. Self-contained blowers contribute to the preservation of the shape of the envelope, the blowers and the ballonets being under the control of the pilot. Planes resembling Venetian blinds facilitate vertical steering, while the suspension of the keel is carried out in such a manner as to secure uniformity of weight upon the gas bag. The propelling power comprises two sets of internal combustion engines, each developing 130 horse-power, the transmission being through rubber belting. The propellers, built of wood, make 350 revolutions per minute, and are set as closely as possible to the centre of resistance.

But the most salient characteristic of this machine is its portability. It can be dismantled and transported by wagons to any desired spot, the suspension frame being constructed in units, each of which is sufficiently small to be accommodated in an ordinary vehicle. Upon arrival the parts may be put together speedily and easily. The authorities submitted the airship to exacting trials and were so impressed by its characteristics and the claims of the inventor that undoubtedly it will be brought into service during the present crisis.

At the same time the whole faith of the German military staff so far as airship operations are concerned, is pinned to the Zeppelin. Notwithstanding its many drawbacks it is the vessel which will be used for the invasion of Great Britain. Even the harbour question, which is admitted to be somewhat acute, has been solved to a certain degree. At strategical points permanent harbours or airship sheds have been established. Seeing that the airships demand considerable skill in docking and undocking, and that it is impossible to achieve these operations against the wind, swinging sheds have been adopted.

On water the practice is to anchor a floating harbour at one end, leaving the structure to swing round with the wind. But on dry land such a dock is impossible. Accordingly turntable sheds have been adopted. The shed is mounted upon a double turn-table, there being two circular tracks the one near the centre of the shed and the other towards its extremities. The shed is mounted upon a centre pivot and wheels engaged with these inner and outer tracks. In this manner the shed may be swung round to the most favourable point of the compass according to the wind.

In the field, however, such practices are impossible, and the issue in this connection has been overcome by recourse to what may be termed portable harbours. They resemble the tents of peripatetic circuses and travelling exhibitions. There is a network of vertical steel members which may be set with facility and speed and which are stayed by means of wire guys. At the top of the outer vertical posts pulleys are provided whereby the outer skin or canvas forming the walls may be hauled into position, while at the apex of the roof further pulleys ensure the proper placing of the roofing. The airship is able to enter or leave from either end according to conditions. The material is fireproofed as a precautionary measure, but at the same time the modern aerial bomb is able to penetrate the roofing without any difficulty and to explode against the airship anchored within.

The one great objection to the Zeppelin harbour is the huge target it offers to hostile attack, which, in the event of a vessel being moored within, is inevitably serious. Thus, for instance, upon the occasion of the air raids conducted by Lieutenant Collet and of Squadron Commander Briggs and his colleagues at Dusseldorf and Friedrichshafen respectively, little difficulty was experienced in destroying the airships riding at anchor. The target offered by the shed is so extensive that it would be scarcely possible for a flying enemy to miss it. A bomb dropped from a reasonable height, say 500 feet, would be almost certain to strike some part of the building, and a Zeppelin is an easy vessel to destroy. The firing of one balloon is sufficient to detonate the whole, for the simple reason that hydrogen gas is continuously oozing through the bags in which it is contained. According to a recent statement the Germans are said to be utilising an inert or non-inflammable gas, equal in lifting power to hydrogen, for the inflation of military craft, but scientific thought does not entertain this statement with any degree of seriousness. No gas as light as hydrogen and non-explosive is known to commerce.

Will Germany invade Great Britain by air? This is the absorbing topic of the moment--one which has created intense interest and a certain feeling of alarm among the timorous. Although sporadic raids are considered to be possible and likely to be carried out with a varying measure of success--such as that made upon the British East Coast--eminent authorities ridicule an invasion in force. The risk would be enormous, although there is no doubt that Germany, which has always maintained that an invasion of this character will be made, will be compelled to essay such a task, in order to satisfy public opinion, and to justify official statements. It is a moot point, however, whether the invaders ever will succeed in making good their escape, unless Nature proves exceptionally kind.

The situation is best summed up in the unbiassed report of General George P. Scriven, Chief Signal officer of the United States Army to the U.S. Secretary of War. In this report, which deals exhaustively with the history, construction and achievements of airships, such an invasion is described as fantastic and impracticable. Writing on November 10th, 1914, the officer declares that "he is not prepared to recommend the American Army to take up seriously the question of constructing dirigibles, as they are not worth their cost as offensive machines, while for reconnaissance or defence they are of far less value than aeroplanes." In his words, "Dirigibles are seemingly useless in defence against the aeroplane or gun-fire."

In order to be able to make an invasion in force upon Great Britain's cities extremely favourable weather must prevail, and the treacherous nature of the weather conditions of the North Sea are known fully well both to British and Teuton navigators. Seeing that the majority of the Zeppelin pilots are drawn from the Navy and mercantile marine, and thus are conversant with the peculiarities and characteristics of this stretch of salt water, it is only logical to suppose that their knowledge will exert a powerful influence in any such decision, the recommendations of the meteorological savants not withstanding.

When the Zeppelin pride of the German Navy "L-1" was hurled to destruction by a typical North Sea squall, Captain Blew of the Victoria Luise, a Zeppelin with many great achievements to her credit, whose navigator was formerly in the Navy, and thus is familiar with the whole issue, explained that this atmospheric liveliness of the North Sea prevails for the most part in the latitude of Norway, but that it frequently extends as far south as the gate of the Channel. He related furthermore that the rain squalls are of tropical violence, while the vertical thrusts of air are such that no dirigible as yet constructed could ever hope to live in them. Under such conditions, he continued, the gas is certain to cool intensely, and the hull must then become waterlogged, not to mention the downward thrust of the rain. Under such conditions buoyancy must be imperilled to such a degree as to demand the jettisoning of every piece of ballast, fuel and other removable weight, including even the steadying and vertical planes. When this has been done, he pointed out, nothing is left with which to combat the upward vertical thrusts of the air. To attempt to run before the wind is to court positive disaster, as the wind is certain to gain the mastery. Once the airship loses steering way and is rendered uncontrollableit becomes the sport of the forces of Nature, with the result that destruction is merely a matter of minutes, or even seconds.

Every navigator who knows the North Sea will support these conclusions.

Squalls and blizzards in winter, and thunderstorms in summer, rise with startling suddenness and rage with terrific destructive fury. Such conditions must react against the attempt of an aerial invasion in force, unless it be made in the character of the last throw by a desperate gambler, with good fortune favouring the dash to a certain degree. But lesser and more insignificant Zeppelin raids are likely to be somewhat frequent, and to be made at every favourable climatic opportunity.

CHAPTER VII

AEROPLANES OF WAR

Owing to the fertility of inventors and the resultant multiplicity of designs it is impossible to describe every type of heavier-than-air machine which has been submitted to the exacting requirements of military duty. The variety is infinite and the salient fact has already been established that many of the models which have proved reliable and efficient under normal conditions are unsuited to military operations. The early days of the war enabled those of doubtful value to be eliminated, the result being that those machines which are now in use represent the survival of the fittest. Experience has furthermore emphasised the necessity of reducing the number of types to the absolute minimum. This weeding-out process is being continued and there is no doubt that by the time the war is concluded the number of approved types of aeroplanes of military value will have been reduced to a score or less. The inconveniences and disadvantages arising from the utilisation of a wide variety of different types are manifold, the greatest being the necessity of carrying a varied assortment of spare parts, and confusion in the repair and overhauling shops.

The methodical Teuton was the first to grasp the significance of these drawbacks; he has accordingly carried standardisation to a high degree of efficiency, as is shown in another chapter. At a later date France appreciated the wisdom of the German practice, and within a short time after the outbreak of hostilities promptly ruled out certain types of machines which were regarded as unsuitable. In this instance the process of elimination created considerable surprise, inasmuch as it involved an embargo on the use of certain machines, which under peace conditions had achieved an international reputation, and were held to represent the finest expression of aeronautical science in France as far as aeroplane developments are concerned.

Possibly the German machine which is most familiar, by name, to the general public is the Taube, or, as it is sometimes called, the Etrich monoplane, from the circumstance that it was evolved by the Austrian engineer Igo Etrich in collaboration with his colleague Wels. These two experimenters embarked on the study of dynamic flight contemporaneously with Maxim, Langley, Kress, and many other well-known pioneers, but it was not until 1908 that their first practical machine was completed. Its success was instantaneous, many notable flights being placed to its credit, while some idea of the perfection of its design may be gathered from the fact that the machine of to-day is substantially identical with that used seven years ago, the alterations which have been effected meanwhile being merely modifications in minor details.

The design of this machine follows very closely the lines of a bird in flight-- hence its colloquial description, "Taube," or "dove." Indeed the analogy to the bird is so close that the ribs of the frame resemble the feathers of a bird. The supporting plane is shaped in the manner of a bird's distended wing, and is tipped up at the rear ends to ensure stability. The tail also resembles that of a bird very closely.

This aeroplane, especially the latest type, is very speedy, and it has proved extremely reliable. It is very sharp in turning and extremely sensitive to its rudder, which renders it a first-class craft for reconnoitring duty. The latest machines are fitted with motors developing from 120 to 150 horse-power.

The "Taube" commanded attention in Germany for the reason that it indicated the first departure from the adherence to the French designs which up to that time had been followed somewhat slavishly, owing to the absence of native initiative.

The individuality of character revealed in the "Taube" appealed to the German instinct, with the result that the machine achieved a greater reputation than might have been the case had it been pitted against other types of essentially Teutonic origin. The Taube was subsequently tested both in France and Great Britain, but failed to raise an equal degree of enthusiasm, owing to the manifestation of certain defects which marred its utility. This practical experience tended to prove that the Taube, like the Zeppelin, possessed a local reputation somewhat of the paper type. The Germans, however, were by no means disappointed by such adverse criticism, but promptly set to work to eliminate defects with a view to securing an all-round improvement.

The most successful of these endeavours is represented in the Taube-Rumpler aeroplane, which may be described as an improved edition of Etrich's original idea. As a matter of fact the modifications were of so slight, though important, a character that many machines generically described as Taubes are in reality Rumplers, but the difference is beyond detection by the ordinary and unpractised observer.

In the Rumpler machine the wings, like those of the Taube, assume broadly the form and shape of those of the pigeon or dove in flight. The early Rumpler machines suffered from sluggish control, but in the later types this defect has been overcome. In the early models the wings were flexible, but in the present craft they are rigid, although fitted with tips or ailerons. The supporting truss beneath the wings, which was such an outstanding feature of its prototype, has been dispensed with, the usual I-beam longitudinals being used in its stead. The latest machines fitted with 100-120 horse-power Mercedes motors have a fine turn of speed, possess an enhanced ascensional effort, and are far simpler to control

Other German machines which are used in the military service are the Gotha and the Albatross. The former is a monoplane, and here again the influence of Etrich upon German aeroplane developments is strongly manifested, the shape of the bird's wing being retained. In the Gotha the truss which Etrich introduced is a prominent characteristic. The Albatross is a biplane, but this craft has proved to be somewhat slow and may be said to be confined to what might be described as the heavier aerial military duties, where great endurance and reliability are essential. As the war proceeds, doubtless Teuton ingenuity will be responsible for the appearance of new types, as well as certain modifications in the detail construction of the existing machines, but there is every indication that the broad lines of Etrich's conception will be retained in all monoplanes.

There is one point in which Germany has excelled. Wood is not employed in the construction of these heavier-than-air craft. Steel and the lighter tough alloys are exclusively used. In this way the minimum of weight consistent with the maximum of strength policy is carried out. Moreover the manufacture of component parts is facilitated and accelerated to a remarkable degree by the use of metal, while the tasks of fitting and repairing are notably expedited by the practice of standardisation. Germany is also manifesting commendable enterprise in the perfection of light powerful motors for these dynamic machines. The latest types of explosion-motors range from 100 to 150 horse-power; the advantages of these are obvious.

Upon the outbreak of hostilities the French possessed an enormous number and variety of aeroplanes and this aerial fleet had been brought to a high standard of organisation. The aerial fleet is sub-divided into squadrons called "escadrilles," each of which comprises six machines and pilots. These units are kept up to strength, wastage being made up from reserves, so as to maintain the requisite homogeneity.

But ere the war had been in progress many weeks an official order was issued forbidding the employment of the Bleriot, Deperdussin, Nieuport, and R.E.P. monoplanes. Those which received official approval included the Caudron, Henry, and Maurice Farman, Morane-Saulnier, and Voisin machines.

This drastic order came somewhat as a thunderbolt, and the reason for the decree has not been satisfactorily revealed. Suffice to say that in one stroke the efficiency and numerical strength of the French aerial navy were reduced very appreciably. For instance, it is stated that there were thirty escadrilles of Bleriot monoplanes together with pilots at the front, in addition to thirty mixed escadrilles of the other prohibited types with their fliers. Moreover a round 33 escadrilles of all the various types were in reserve. The effect of the military order was to reduce the effective strength by no fewer than 558 aeroplanes.

Seeing that the French aerial force was placed at a great disadvantage numerically by this action, there seems to be ample justification for the hostile criticism which the decree of prohibition aroused in certain circles, especially when it is remembered that there was not an equal number of the accepted machines available to take the place of those which had been ruled out of court. One effect of this decree was to throw some 400 expert aviators upon the waiting list for the simple reason that machines were unavailable. Some of the best aviation skill and knowledge which France possesses were affected by the order. It is stated that accomplished aviators, such as Vedrines, were unable to obtain machines.

It will be seen that the ultimate effect of the French military decree was to reduce the number of types to about four, each of which was allotted a specific duty. But whereas three different bi-planes are on the approved list there is only one monoplane-- the Morane-Saulaier. This machine, however, has a great turn of speed, and it is also able to climb at a very fast pace. In these respects it is superior to the crack craft of Germany, so that time after time the latter have refused battle in the skies, and have hurried back to their lines.

The Morane-Saulnier is the French mosquito craft of the air and like the insect, it is avowedly aggressive. In fact, its duties are confined to the work of chasing and bringing down the enemy, for which work its high manoeuvring capacity is excellently adapted. Its aggressive armament comprises a mitrailleuse. Unfortunately, however, the factory responsible for the production of this machine is at present handicapped by the limitations of its manufacturing plant, which when pushed to the utmost extent cannot turn out more than about ten machines per week. No doubt this deficiency will be remedied as the war proceeds by extension of the works or by allotting orders to other establishments, but at the time of the decree the manufacturing capacity was scarcely sufficient to make good the wastage, which was somewhat heavy.

As far as biplanes are concerned the Caudron is the fastest in flight and is likewise extremely quick in manoeuvring. It is a very small machine and is extremely light, but the fact that it can climb at the rate of over 330 feet per minute is a distinct advantage in its favour. It supplements the Morane-Saulnier monoplane in the specific duty of the latter, while it is also employed for discovering the enemy's artillery and communicating the range of the latter to the French and British artillery. In this latter work it has played a very prominent part and to it is due in no small measure that deadly accuracy of the artillery of the Allies which has now become so famous. This applies especially to those tactics, where the field artillery dashes up to a position, discharges a number of rounds in rapid succession, or indulges in rafale firing, and then limbering up, rushes away before the enemy can reply.

As is well known the Farman biplanes possess high endurance qualities. They can remain aloft for many hours at a stretch and are remarkably reliable. Owing to these qualities they are utilised for prolonged and searching reconnoitring duties such as strategical reconnaissances as distinct from the hurried and tactical reconnaissances carried out by fleeter machines. While they are not so speedy as the monoplanes of the German military establishment, endurance in this instance is preferable to pace. A thorough survey of the enemy's position over the whole of his military zone, which stretches back for a distance of 30 miles or so from the outer line of trenches, is of incalculable value to a commander who is contemplating any decisive movement or who is somewhat in doubt as to the precise character of his antagonist's tactics.

The French aerial fleet has been particularly active in its work of raiding hostile positions and submitting them to a fusillade of bombs from the clouds. The machine which is allotted this specific task is the Voisin biplane. This is due to the fact that this machine is able to carry a great weight. It was speedily discovered that in bomb-raids it is essential for an aeroplane to be able to carry a somewhat large supply of missiles, owing to the high percentage of misses which attends these operations. A raid by a machine capable of carrying only, say, half-a-dozen projectiles, is virtually a waste of fuel, and the endurance limitations of the fast machines reacts against their profitable use in this work. On the other hand, the fact that the Voisin machine is able to carry a large supply of bombs renders it an ideal craft for this purpose; hence the official decision to confine it to this work.

So far as the British efforts in aerial work are concerned there is no such display of rigid selection as characterises the practice of the French and German military authorities. Britain's position in the air has been extensively due to private enterprise, and this is still being encouraged. Moreover at the beginning of the war Britain was numerically far inferior both to her antagonist and to her ally. Consequently it was a wise move to encourage the private manufacture of machines which had already established their value. The consequence is that a variety of machines figure in the British aerial navy. Private initiative is excellently seconded by the Government manufacturing aeroplane factory, while the training of pilots is likewise being carried out upon a comprehensive scale. British manufacture may be divided into two broad classes--the production of aeroplanes and of waterplanes respectively. Although there is a diversity of types there is a conspicuous homogeneity for the most part, as was evidenced by the British raid carried out on February 11-12, when a fleet of 34 machines raided the various German military centres established along the coast of Flanders.

Considerable secrecy has been displayed by the British Government concerning the types of machines that are being utilised, although ample evidence exists from the producing activity of the various establishments that all available types which have demonstrated their reliability and efficiency are being turned to useful purpose. The Avro and Sopwith warplanes with their very high speeds have proved remarkably successful.

So far as manufacturing is concerned the Royal Aerial Factory may be said to constitute the back bone of the British aerial fleet. This factory fulfils various purposes. It is not only engaged in the manufacture of machines, and the development of aeroplanes for specific duties, but also carries out the inspection and testing of machines built by private firms. Every machine is submitted to an exacting test before it is passed into the service.

Three broad types of Government machines are manufactured at this establishment. There is that designed essentially for scouting operations, in which speed is the all-important factor and which is of the tractor type. Another is the "Reconnoitring" machine known officially as the "R.E." to-day, but formerly as the "B.E" (Bleriot-Experimental), a considerable number of which are in commission.

This machine is also of the tractor type, carrying a pilot and an observer, and has a maximum speed of 40-50 miles per hour. If required it can further be fitted with an automatic gun for defence and attack. The third craft is essentially a fighting machine. Owing to the introduction of the machine-gun which is fixed in the prow, with the marksman immediately behind it, the screw is placed at the rear. The pilot has his seat behind the gunner. The outstanding feature of these machines is the high factor of safety, which attribute has astonished some of the foremost aviation experts in the world.

Great Britain lagged behind her Continental rivals in the development of the Fourth Arm, especially in matters pertaining to motive power. For some time reliance was placed upon foreign light highspeed explosion motors, but private enterprise was encouraged, with the result that British Motors comparing favourably in every respect with the best productions upon the Continent are now available. Development is still proceeding, and there is every evidence that in the near future entire reliance will be placed upon the native motor.

Undoubtedly, as the war progresses, many valuable lessons will be learned which will exercise an important bearing upon the design and construction of warplanes. The ordeals to which the machines are submitted in military duties are far more severe than any imposed by the conditions of commerce. Accordingly there is every indication that the conflict upon the Continent will represent a distinctive epoch in aeroplane design and construction. Many problems still await solution, such as the capacity to hover over a position, and it is quite possible that these complex and baffling questions will be settled definitely as the result of operations in the field. The aeroplane has reached a certain stage of evolution: further progress is virtually impossible unless something revolutionary is revealed, perfected, and brought to the practical stage.

CHAPTER VIII

SCOUTING FROM THE SKIES

From the moment when human flight was lifted from the rut of experiment to the field of practical application, many theories, interesting and illuminating, concerning the utility of the Fourth Arm as a military unit were advanced. The general consensus of expert opinion was that the flying machine would be useful to glean information concerning the movements of an enemy, rather than as a weapon of offence.

The war is substantiating this argument very completely. Although bomb-dropping is practised somewhat extensively, the results achieved are rather moral than material in their effects. Here and there startling successes have been recorded especially upon the British side, but these triumphs are outnumbered by the failures in this direction, and merely serve to emphasise the views of the theorists.

The argument was also advanced that, in this particular work, the aeroplane would prove more valuable than the dirigible, but actual campaigning has proved conclusively that the dirigible and the heavier-than-air machines have their respective fields of utility in the capacity of scouts. In fact in the very earliest days of the war, the British airships, though small and slow in movement, proved more serviceable for this duty than their dynamic consorts. This result was probably due to the fact that military strategy and tactics were somewhat nonplussed by the appearance of this new factor. At the time it was an entirely unknown quantity. It is true that aircraft had been employed in the Balkan and the Italo-Ottoman campaigns, but upon such a limited scale as to afford no comprehensive idea of their military value and possibilities.

The belligerents, therefore, were caught somewhat at a disadvantage, and an appreciable period of time elapsed before the significance of the aerial force could be appreciated, while means of counter acting or nullifying its influences had to be evolved simultaneously, and according to the exigencies of the moment. At all events, the protagonists were somewhat loth to utilise the dirigible upon an elaborate scale or in an aggressive manner. It was employed more after the fashion of a captive balloon, being sent aloft from a point well behind the front lines of the force to which it was attached, and well out of the range of hostile guns. Its manoeuvres were somewhat circumscribed, and were carried out at a safe distance from the enemy, dependence being placed upon the advantages of an elevated position for the gathering of information.

But as the campaign progressed, the airships became more daring. Their ability to soar to a great height offered them complete protection against gun-fire, and accordingly sallies over the hostile lines were carried out. But even here a certain hesitancy became manifest. This was perfectly excusable, for the simple reason that the dirigible, above all, is a fair-weather craft, and disasters, which had overtaken these vessels time after time, rendered prudence imperative. Moreover, but little was known of the range and destructiveness of anti-aircraft guns.

In the duty of reconnoitring the dirigible possesses one great advantage over its heavier-than-air rival. It can remain virtually stationary in the air, the propellers revolving at just sufficient speed to off-set the wind and tendencies to drift. In other words, it has the power of hovering over a position, thereby enabling the observers to complete their task carefully and with deliberation.

On the other hand, the means of enabling an aeroplane to hover still remain to be discovered. It must travel at a certain speed through the air to maintain its dynamic equilibrium, and this speed is often too high to enable the airman to complete his reconnaissance with sufficient accuracy to be of value to the forces below. All that the aeroplane can do is to circle above a certain position until the observer is satisfied with the data he has collected.

But hovering on the part of the dirigible is not without conspicuous drawbacks. The work of observation cannot be conducted with any degree of accuracy at an excessive altitude. Experience has proved that the range of the latest types of anti- aircraft weapons is in excess of anticipations. The result is that the airship is useless when hovering beyond the zone of fire. The atmospheric haze, even in the clearest weather, obstructs the observer's vision. The caprices of this obstacle are extraordinary, as anyone who has indulged in ballooning knows fully well. On a clear summer's day I have been able to see the ground beneath with perfect distinctness from a height of 4,500 feet, yet when the craft had ascended a further two or three hundred feet, the panorama was blurred. A film of haze lies between the balloon and the ground beneath. And the character of this haze is continually changing, so that the aerial observer's task is rendered additionally difficult. Its effects are particularly notice able when one attempts to photograph the view unfolded below. Plate after plate may be exposed and nothing will be revealed. Yet at a slightly lower altitude the plates may be exposed and perfectly sharp and well-defined images will be obtained.

Seeing that the photographic eye is keener and more searching than the human organ of sight, it is obvious that this haze constitutes a very formidable obstacle. German military observers, who have accompanied the Zeppelins and Parsevals on numerous aerial journeys under varying conditions of weather, have repeatedly drawn attention to this factor and its caprices, and have not hesitated to venture the opinion that it would interfere seriously with military aerial reconnaissances, and also that it would tend to render such work extremely hazardous at times.

When these conditions prevail the dirigible must carry out its work upon the broad lines of the aeroplane. It must descend to the level where a clear view of the ground may be obtained, and in the interests of safety it has to keep on the move. To attempt to hover within 4,000 feet of the ground is to court certain disaster, inasmuch as the vessel offers a magnificent and steady target which the average gunner, equipped with the latest sighting devices and the most recent types of guns, scarcely could fail to hit.

But the airman in the aeroplane is able to descend to a comparatively low level in safety. The speed and mobility of his machine constitute his protection. He can vary his altitude, perhaps only thirty or forty feet, with ease and rapidity, and this erratic movement is more than sufficient to perplex the marksmen below, although the airman is endangered if a rafale is fired in such a manner as to cover a wide zone.

Although the aeroplane may travel rapidly it is not too fleet for a keen observer who is skilled in his peculiar task. He may only gather a rough idea of the disposition of troops, their movements, the lines of communication, and other details which are indispensable to his commander, but in the main the intelligence will be fairly accurate. Undulating flight enables him to determine speedily the altitude at which he is able to obtain the clearest views of the country beneath. Moreover, owing to his speed he is able to complete his task in far less time than his colleague operating in the dirigible, the result being that the information placed at the disposal of his superior officers is more to the moment, and accordingly of greater value.

Reconnoitring by aeroplane may be divided into two broad categories, which, though correlated to a certain degree, are distinctive, because each constitutes a specific phase in military operations. They are known respectively as "tactical" and "strategical" movements. The first is somewhat limited in its scope as compared with the latter, and has invariably to be carried out rapidly, whereas the strategical reconnaissance may occupy several hours.

The tactical reconnaissance concerns the corps or divisional commander to which the warplane is attached, and consequently its task is confined to the observation of the line immediately facing the particular corps or division. The aviator does not necessarily penetrate beyond the lines of the enemy, but, as a rule limits his flight to some distance from his outermost defences. The airman must possess a quick eye, because his especial duty is to note the disposition of the troops immediately facing him, the placing of the artillery, and any local movements of the forces that may be in progress. Consequently the aviator engaged on this service may be absent from his lines for only a few minutes, comparatively speaking; the intelligence he acquires must be speedily communicated to the force to which he is attached, because it may influence a local movement.

The strategical reconnaissance, on the other hand, affects the whole plan of campaign. The aviators told off for this duty are attached to the staff of the Commander-in-Chief, and the work has to be carried out upon a far more comprehensive and elaborate scale, while the airmen are called upon to penetrate well into the hostile territory to a point thirty, forty, or more miles beyond the outposts.

The procedure is to instruct the flier either to carry out his observations of the territory generally, or to report at length upon a specified stretch of country. In the latter event he may fly to and fro over the area in question until he has acquired all the data it is possible to collect. His work not only comprises the general disposition of troops, defences, placing of artillery, points where reserves are being held, high-roads, railways, base camps, and so forth, but he is also instructed to bring back as correct an idea as possible of what the enemy proposes to do, so that his Commander-in-Chief may adjust his moves accordingly. In order to perform this task with the requisite degree of thoroughness it is often necessary for the airman to remain in the air for several hours continuously, not returning, in fact, until he has completed the allotted duty.

The airman engaged in strategical aerial reconnaissance must possess, above all things, what is known as a "military" eye concerning the country he traverses. He must form tolerably correct estimates of the forces beneath and their character. He must possess the ability to read a map rapidly as he moves through the air and to note upon it all information which is likely to be of service to the General Staff. The ability to prepare military sketches rapidly and intelligibly is a valuable attribute, and skill in aerial photography is a decidedly useful acquisition.

Such men must be of considerable stamina, inasmuch as great demands are made upon their powers of endurance. Being aloft for several hours imposes a severe tax upon the nervous system, while it must also be borne in mind that all sorts and conditions of weather are likely to be encountered, more particularly during the winter. Hail, rain, and blizzards may be experienced in turn, while the extreme cold which often prevails in the higher altitudes during the winter season is a fearful enemy to combat. Often an airman upon his return from such a reconnaissance has been discovered to be so numbed and dazed as a result of the prolonged exposure, that considerable time has elapsed before he has been sufficiently restored to set forth the results of his observations in a coherent, intelligible manner for the benefit of the General Staff. Under these circumstances it is not surprising that the most skilful and experienced aviators are generally reserved for this particular work. In addition to the natural accidents to which the strategical aerial observer is exposed, the dangers arising from hostile gun-fire must not be overlooked. He is manoeuvring the whole time over the enemy's firing zone, where anti-aircraft weapons are disposed strategically, and where every effort is made by artillery to bring him down, or compel him to repair to such a height as to render observation with any degree of accuracy well-nigh impossible.

The methods practised by the German aerial scout vary widely, and are governed in no small measure by the intrepidity and skill of the airman himself. One practice is to proceed alone upon long flights over the enemy's lines, penetrating just as far into hostile territory as the pilot considers advisable, and keeping, of course, within the limits of the radius of action of the machine, as represented by the fuel supply, the while carefully taking mental stock of all that he observes below. It is a kind of roving commission without any definite aim in view beyond the collection of general intelligence.

This work, while productive and valuable to a certain degree, is attended with grave danger, as the German airmen have repeatedly found to their cost. Success is influenced very materially by the accuracy of the airman's judgment. A slight miscalculation of the velocity and direction of the wind, or failure to detect any variations in the climatic conditions, is sufficient to prove his undoing. German airmen who essayed journeys of discovery in this manner, often failed to regain their lines because they ventured too far, misjudged the speed of the wind which was following them on the outward run, and ultimately were forced to earth owing to the exhaustion of the fuel supply during the homeward trip; the increased task imposed upon the motor, which had to battle hard to make headway, caused the fuel consumption per mile to exceed calculations.

Then the venturesome airman cannot neglect another factor which is adverse to his success. Hostile airmen lie in wait, and a fleet of aeroplanes is kept ready for instant service. They permit the invader to penetrate well into their territory and then ascend behind him to cut off his retreat. True, the invader has the advantage of being on the wing, while the ether is wide and deep, without any defined channels of communication. But nine times out of ten the adventurous scout is trapped. His chances of escape are slender, because his antagonists dispose themselves strategically in the air. The invader outpaces one, but in so doing comes within range of another. He is so harassed that he either has to give fight, or, finding his retreat hopelessly cut off, he makes a determined dash, trusting to his high speed to carry him to safety. In these driving tactics the French and British airmen have proved themselves adepts, more particularly the latter, as the chase appeals to their sporting instincts. There is nothing so exhilarating as a quarry who displays a determination to run the gauntlet.

The roving Teuton scout was considerably in evidence in the early days of the war, but two or three weeks' experience emphasised the sad fact that, in aerial strategy, he was hopelessly outmatched by his opponents. His advantage of speed was nullified by the superior tactical and strategical acumen of his antagonists, the result being that the German airman, who has merely been trained along certain lines, who is in many cases nothing more than a cog-wheel in a machine, and who is proverbially slow-witted, has concluded that he is no match for the airmen of the Allies. He found from bitter experience that nothing afforded the Anglo-French military aviators such keen delight as to lie in wait for a "rover," and then to swoop into the air to round him up.

The proportion of these individual scouts who were either brought down, or only just succeeded in reaching safety within their own lines, and who were able to exhibit serious wounds as evidence of the severity of the aerial tussle, or the narrowness of the escape, has unnerved the Teuton airmen as a body to a very considerable extent. Often, even when an aeroplane descended within the German lines, it was found that the roving airman had paid the penalty for his rashness with his life, so that his journey had proved in vain, because all the intelligence he had gained had died with him, or, if committed to paper, was so unintelligible as to prove useless.

It was the success of the British airmen in this particular field of duty which was responsible for the momentous declaration in Field-Marshal Sir John French's famous despatch:--"The British Flying Corps has succeeded in establishing an individual ascendancy, which is as serviceable to us as it is damaging to the enemy . . . . The enemy have been less enterprising in their flights. Something in the direction of the mastery of the air has already been gained."

The methods of the British airmen are in vivid contrast to the practice of the venturesome Teuton aerial rovers described above. While individual flights are undertaken they are not of unknown duration or mileage. The man is given a definite duty to perform and he ascends merely to fulfil it, returning with the information at the earliest possible moment. It is aerial scouting with a method. The intelligence is required and obtained for a specific purpose, to govern a contemplated move in the grim game of war.

Even then the flight is often undertaken by two or more airmen for the purpose of checking and counterchecking information gained, or to ensure such data being brought back to headquarters, since it is quite possible that one of the party may fall a victim to hostile fire. By operating upon these lines there is very little likelihood of the mission proving a complete failure. Even when raids upon certain places such as Dusseldorf, Friedrichshafen or Cuxhaven are planned, complete dependence is not placed on one individual. The machine is accompanied, so that the possibility of the appointed task being consummated is transformed almost into a certainty.

The French flying men work upon broadly similar lines. Their fleet is divided into small squadrons each numbering four, six, or more machines, according to the nature of the contemplated task. Each airman is given an area of territory which is to be reconnoitred thoroughly. In this way perhaps one hundred or more miles of the enemy's front are searched for information at one and the same time. The units of the squadron start out, each taking the appointed direction according to the preconceived plan, and each steering by the aid of compass and map. They are urged to complete the work with all speed and to return to a secret rendezvous.

Later the air is alive with the whirring of motors. The machines are coming back and all converging to one point. They vol-plane to the earth and gracefully settle down within a short distance of each other at the rendezvous. The pilots collect and each relates the intelligence he has gained. The data are collated and in this manner the General Staff is able to learn exactly what is transpiring over a long stretch of the hostile lines, and a considerable distance to the rear of his advance works. Possibly five hundred square miles have been reconnoitred in this manner. Troops have been massed here, lines of communication extend somewhere else, while convoys are moving at a third place. But all has been observed, and the commanding officer is in a position to re-arrange his forces accordingly. It is a remarkable example of method in military tactics and strategy, and conveys a striking idea of the degree to which aerial operations have been organised.

After due deliberation it is decided that the convoys shall be raided, or that massed troops shall be thrown into confusion, if not dispersed. The squadron is ordered to prepare for another aerial journey. The roads along which the convoys are moving are indicated upon the map, or the position of the massed troops in bivouac is similarly shown. The airmen load their machines with a full charge of bombs. When all is ready the leader ascends, followed in rapid succession by the other units, and they whirr through the air in single file. It now becomes a grim game of follow-my-leader.

The leader detects the convoy, swoops down, suddenly launches his missiles, and re-ascends. He does not deviate a foot from his path to observe the effects of his discharge, as the succeeding aeroplane is close behind him. If the leader has missed then the next airman may correct his error. One after another the machines repeat the manoeuvre, in precisely the same manner as the units of a battleship squadron emulate the leading vessel when attacking the foe. The tactical evolutions have been laid down, and there is rigid adherence thereto, because only thereby may success be achieved. When the last war-plane has completed its work, the leader swings round and repeats the dash upon the foe. A hail of bullets may scream around the men in the air, but one and all follow faithfully in the leader's trail. One or more machines may fail in the attack, and may even meet with disaster, but nothing interferes with the movements of the squadron as a whole. It is the homogeneity of the attacking fleet which tells, and which undermines the moral of the enemy, even if it does not wreak decisive material devastation. The work accomplished to the best of their ability, the airmen speed back to their lines in the same formation.

At first sight reconnoitring from aloft may appear a simple operation, but a little reflection will reveal the difficulties and arduousness of the work. The observer, whether he be specially deputed, or whether the work be placed in the hand of the pilot himself--in this event the operation is rendered additionally trying, as he also has to attend to his machine must keep his eyes glued to the ground beneath and at the same time be able to read the configuration of the panorama revealed to him. He must also keep in touch with his map and compass, so as to be positive of his position and direction. He must be a first-class judge of distances and heights.

When flying rapidly at a height of 4,000 feet or more, the country below appears as a perfect plane, or flat stretch, although as a matter of fact it may be extremely undulating. Consequently, it is by no means a simple matter to distinguish eminences and depressions, or to determine the respective and relative heights of hills.

If a rough sketch is required, the observer must be rapid in thought, quick in determination, and facile with his pencil, as the machine, no matter how it may be slowed down, is moving at a relatively high speed. He must consult his map and compass frequently, since an airman who loses his bearings is useless to his commander-in-chief. He must have an eagle eye, so as to be able to search the country unfolded below, in order to gather all the information which is likely to be of value to his superior officers. He must be able to judge accurately the numbers of troops arrayed beneath him, the lines of the defensive works, to distinguish the defended from the dummy lines which are thrown up to baffle him, and to detect instantly the movement of the troops and the direction, as well as the roads, along which they are proceeding. Reserves and their complement, artillery, railway-lines, roads, and bridges, if any, over streams and railways must be noted--in short he must obtain an eye photograph of the country he observes and grasp exactly what is happening there. In winter, with the thermometer well down, a blood-freezing wind blowing, wreaths of clouds drifting below and obscuring vision for minutes at a time, the rain possibly pelting down as if presaging a second deluge, the plight of the vigilant human eye aloft is far from enviable.

Upon the return of the machine to its base, the report must be prepared without delay. The picture recorded by the eye has to be set down clearly and intelligibly with the utmost speed. The requisite indications must be made accurately upon the map. Nothing of importance must be omitted: the most trivial detail is often of vital importance.

A facile pencil is of inestimable value in such operations. While aloft the observer does not trust to his memory or his eye picture, but commits the essential factors to paper in the form of a code, or what may perhaps be described more accurately as a shorthand pictorial interpretation of the things he has witnessed. To the man in the street such a record would be unintelligible, but it is pregnant with meaning, and when worked out for the guidance of the superior officers is a mass of invaluable detail.

At times it so happens that the airman has not been able to complete his duty within the time anticipated by those below. But he has gathered certain information which he wishes to communicate without coming to earth. Such data may be dropped from the clouds in the form of maps or messages. Although wireless telegraphy is available for this purpose, it suffers from certain drawbacks. If the enemy possesses an equipment which is within range of that of the air-craft and the force to which it belongs, communications may be nullified by the enemy throwing out a continuous stream of useless signals which "jamb" the intelligence of their opponents.

If a message--written in code--or a map is to be dropped from aloft it is enclosed within a special metallic cylinder, fitted with a vane tail to ensure direction of flight when launched, and with a detonating head. This is dropped overboard. When it strikes the ground the detonator fires a charge which emits a report without damaging the message container, and at the same time fires a combustible charge emitting considerable smoke. The noise attracts anyone in the vicinity of the spot where the message has fallen, while at the same time the clouds of smoke guide one to the point and enable the cylinder to be recovered. This device is extensively used by the German aviators, and has proved highly serviceable; a similar contrivance is adopted by French airmen.

There is one phase of aerial activity which remains to be demonstrated. This is the utilisation of aerial craft by the defenders of a besieged position such as a ring of fortifications or fortified city. The utility of the Fourth Arm in this province has been the subject of considerable speculation. Expert opinion maintains that the advantage in this particular connection would rest with the besiegers. The latter would be able to ascertain the character of the defences and the defending gun-force, by means of the aerial scout, who would prove of inestimable value in directing the fire of the besieging forces.

On the other hand it is maintained that an aerial fleet would be useless to the beleaguered. In the first place the latter would experience grave difficulties in ascertaining the positions of the attacking and fortress-reducing artillery, inasmuch as this could be masked effectively, and it is thought that the aerial force of the besieged would be speedily reduced to impotence, since it would be subjected to an effective concentrated fire from the ring of besieging anti-aircraft guns and other weapons. In other words, the theory prevails that an aerial fleet, no matter how efficient, would be rendered ineffective for the simple reason that it would be the initial object of the besieger's attack. Possibly the stem test of experience will reveal the fallacy of these contentions as emphatically as it has disproved others. But there is one point upon which authorities are unanimous. If the artillery of the investing forces is exposed and readily distinguishable, the aerial forces of the beleaguered will bring about its speedy annihilation, as the defensive artillery will be concentrated upon that of the besiegers.

CHAPTER IX

THE AIRMAN AND ARTILLERY

There is one field in which the airman has achieved distinctive triumphs. This is in the guidance of artillery fire. The modern battle depends first and foremost upon the fierce effec tiveness of big-gun assault, but to ensure this reliable direction is imperative. No force has proved so invaluable for this purpose as the man of-the-air, and consequently this is the province in which he has been exceptionally and successfully active.

It will be recalled that in the Japanese investiture of Port Arthur during the Russo-Japanese war, thousands of lives were expended upon the retention and assault of 203 Metre Hill. It was the most blood-stained spot upon the whole of the Eastern Asiatic battlefield. General Nogi threw thousands after thousands of his warriors against this rampart while the Russians defended it no less resolutely. It was captured and re-captured; in fact, the fighting round this eminence was so intense that it appeared to the outsider to be more important to both sides than even Port Arthur itself.

Yet if General Nogi had been in the possession of a single aeroplane or dirigible it is safe to assert that scarcely one hundred Japanese or Russian soldiers would have met their fate upon this hill. Its value to the Japanese lay in one sole factor. The Japanese heavy guns shelling the harbour and the fleet it contained were posted upon the further side of this eminence and the fire of these weapons was more or less haphazard. No means of directing the artillery upon the vital points were available; 203 Metre Hill interrupted the line of sight. The Japanese thereupon resolved to capture the hill, while the Russians, equally appreciative of the obstruction it offered to their enemy, as valiantly strove to hold it. Once the hill was captured and the fire of the Japanese guns could be directed, the fate of the fortress was sealed.

Similar conditions have prevailed during the present campaign, especially in the western theatre of war, where the ruggedness of the country has tended to render artillery fire ineffective and expensive unless efficiently controlled. When the German Army attacked the line of the British forces so vehemently and compelled the retreat at Mons, the devastating fire of the enemy's artillery was directed almost exclusively by their airmen, who hovered over the British lines, indicating exactly the point where gun-fire could work the maximum of havoc. The instant concentration of massed artillery fire upon the indicated positions speedily rendered one position after another untenable.

The Germans maintained the upper hand until at last the aerial forces of the British Expeditionary Army came into action. These airmen attacked the Teuton aerial craft without the slightest hesitation, and in a short while rendered cloudland absolutely unhealthy. The sequel was interesting. As if suddenly blinded, the German artillery fire immediately deteriorated. On the other hand, the British artillery, now having the benefit of aerial guidance, was able to repay the German onslaughts with interest, and speedily compelled that elaborate digging-in of the infantry lines which has now become so characteristic of the opposing forces.

So far as the British lines are concerned the men in the trenches keep a sharp look-out for hostile aeroplanes. The moment one is observed to be advancing, all the men seclude themselves and maintain their concealment. To do otherwise is to court a raking artillery outburst. The German aeroplane, detecting the tendency of the trenches describes in the air the location of the vulnerable spot and the precise disposition by flying immediately above the line. Twice the manoeuvre is repeated, the second movement evidently being in the character of a check upon the first observation, and in accordance with instructions, whereupon the Tommies, to quote their own words, "know they are in for it!" Ere the aeroplane has completed the second manoeuvre the German guns ring out.

The facility with which artillery fire can be concentrated through the medium of the aeroplane is amazing. In one instance, according to the story related to me by an officer, "a number of our men were resting in an open field immediately behind the second line of trenches, being in fact the reserves intended for the relief of the front lines during the following night. An aeroplane hove in sight. The men dropped their kits and got under cover in an adjacent wood. The aeroplane was flying at a great height and evidently laboured under the impression that the kits were men. Twice it flew over the field in the usual manner, and then the storm of shrapnel, 'Jack Johnsons' and other tokens from the Kaiser rained upon the confined space. A round four hundred shells were dropped into that field in the short period of ten minutes, and the range was so accurate that no single shell fell outside the space. Had the men not hurried to cover not one would have been left alive to tell the tale, because every square foot of the land was searched through and through. We laughed at the short-sightedness of the airman who had contributed to such a waste of valuable shot and shell, but at the same time appreciated the narrowness of our own escape."

The above instance is by no means isolated. It has happened time after time. The slightest sign of activity in a trench when a "Taube" is overhead suffices to cause the trench to be blown to fragments, and time after time the British soldiers have had to lie prone in their trenches and suffer partial burial as an alternative to being riddled by shrapnel.

The method of ascertaining the range of the target from the indications given by the aeroplane are of the simplest character. The German method is for the aerial craft to fly over the position, and when in vertical line therewith to discharge a handful of tinsel, which, in falling, glitters in the sunlight, or to launch a smoking missile which answers the same purpose as a projectile provided with a tracer. This smoke-ball being dropped over the position leaves a trail of black or whitish smoke according to the climatic conditions which prevail, the object being to enable the signal to be picked up with the greatest facility. The height at which the aerial craft is flying being known, a little triangulation upon the part of the observer at the firing point enables him to calculate the range and to have the guns laid accordingly.

When the aerial craft has been entrusted with the especial duty of directing artillery-fire, a system of communication between the aerial observer and the officer in charge of the artillery is established, conducted, of course, by code. In the British Army, signalling is both visual and audible. In daylight visual signalling is carried out by means of coloured flags or streamers and smoke-signals, while audible communication is effected by means of a powerful horn working upon the siren principle and similar to those used by automobiles. Both flags and sound-signals, however, are restricted owing to the comparatively short distances over which they can be read with any degree of accuracy. The smoke-signal therefore appears to be the most satisfactory and reliable, as the German airmen have proved conclusively, for the simple reason that the trail of smoke may be picked up with comparative ease, even at a distance, by means of field glasses. The tinsel too, is readily distinguishable, particularly in bright weather, for the glittering surface, catching the sun-light, acts some what in the manner of a heliograph.

The progress of the airman is followed by two officers at the base from which he started. One is equipped with the director, while the second takes the range. Directly this has been found as a result of calculation, the guns are laid ready for firing. In those cases where the enemy's artillery is concealed perhaps behind a hill, the airman is of incalculable value, inasmuch as he is able to reveal a position which otherwise would have to be found by considerable haphazard firing, and which, even if followed by a captive balloon anchored above the firing point, might resist correction.

The accuracy of the airman's work in communicating the range has been responsible for the high efficiency of the British and French artillery. The latter, with the 75 millimetre quick-firing gun, is particularly adapted to following up the results of the aeroplane's reconnaissance, especially with the system of rafale fire, because the whole position can be searched through and through within a minute or two. According to information which has been given to me by our artillery officers, the British system also has proved disastrous to the enemy. The practice is to get the range as communicated by the aeroplane, to bring the artillery into position speedily, to discharge salvo after salvo with all speed for a few minutes, and then to wheel the artillery away before any hostile fire can be returned. The celerity with which the British artillery comes into, and goes out of, action has astonished even our own authorities. This mobility is of unique value: it is taking advantage of a somewhat slow-witted enemy with interest. By the time the Germans have opened fire upon the point whence the British guns were discharged, the latter have disappeared and are ready to let fly from another point, some distance away, so that the hostile fire is abortive. Mobility of such a character is decidedly unnerving and baffling even to a quick-witted opponent.

In his search for hostile artillery the airman runs grave risks and displays remarkable resource. It is invariably decided, before he sets out, that he shall always return to a certain altitude to communicate signals. Time after time the guns of the enemy have been concealed so cunningly from aerial observation as to pass unnoticed. This trait became more pronounced as the campaigns of the Aisne progressed. Accordingly the airman adopts a daring procedure. He swoops down over suspicious places, where he thinks guns may be lurking, hoping that the enemy will betray its presence. The ruse is invariably successful. The airman makes a sudden dive towards the earth. The soldiers in hiding below, who have become somewhat demoralised by the accuracy of the British aerial bomb-throwers, have an attack of nerves. They open a spirited fusillade in the hope of bringing the airman to earth. But their very excitement contributes to his safety. The shots are fired without careful aim and expend themselves harmlessly. Sweeping once more upwards, the airman regains the pre-determined level, performs a certain evolution in the air which warns the observer at his base that he has made a discovery, and promptly drops his guiding signal directly over the point from which he has drawn fire.

Operations at night are conducted by means of coloured lights or an electrical searchlight system. In the former instance three lights are generally carried--white, red, and green--each of which has a distinctive meaning. If reliance is placed upon the electric light signalling lamp, then communications are in code. But night operations are somewhat difficult and extremely dangerous, except when the elements are propitious. There is the ground mist which blots everything from sight, rendering reconnaissance purely speculative. But on a clear night the airman is more likely to prove successful. He keeps a vigilant eye upon all ground-lights and by close observation is able to determine their significance. It is for this reason that no lights of any description are permitted in the advance trenches. The striking of a match may easily betray a position to the alert eye above.

So far as the British Army is concerned a complete code is in operation for communicating between aeroplanes and the ground at night. Very's lights are used for this purpose, it being possible to distinguish the respective colours at a distance of six miles and from an altitude of 2,000 feet. The lights are used both by the aeroplane and the battery of artillery.

The code is varied frequently, but the following conveys a rough idea of how communication is carried out by this means under cover of darkness. The aeroplane has located its objective and has returned to the pre-arranged altitude. A red light is thrown by the airman. It indicates that he is directly over the enemy's position. A similarly coloured light is shown by the artillery officer, which intimates to the airman that his signal has been observed and that the range has been taken.

In observing the effects of artillery fire a code of signals is employed between the airman and the artillery officer to indicate whether the shot is "long" or "short," to the right or to the left of the mark, while others intimate whether the fuse is correctly timed or otherwise. It is necessary to change the code fairly frequently, not only lest it should fall into the enemy's hands, but also to baffle the hostile forces; otherwise, after a little experience, the latter would be able to divine the significance of the signals, and, in anticipation of being greeted with a warm fusillade, would complete hurried arrangements to mitigate its effects, if not to vacate the position until the bombardment had ceased.

Sufficient experience has already been gathered, however, to prove the salient fact that the airman is destined to play an important part in the direction and control of artillery-fire. Already he has been responsible for a re-arrangement of strategy and tactics. The man aloft holds such a superior position as to defy subjugation; the alternative is to render his work more difficult, if not absolutely impossible.

## CHAPTER X

## BOMB-THROWING FROM AIR-CRAFT

During the piping times of peace the utility of aircraft as weapons of offence was discussed freely in an academic manner. It was urged that the usefulness of such vessels in this particular field would be restricted to bomb-throwing. So far these contentions have been substantiated during the present campaign. At the same time it was averred that even as a bomb-thrower the ship of the air would prove an uncertain quantity, and that the results achieved would be quite contrary to expectations. Here again theory has been supported by practice, inasmuch as the damage wrought by bombs has been comparatively insignificant.

The Zeppelin raids upon Antwerp and Britain were a fiasco in the military sense. The damage inflicted by the bombs was not at all in proportion to the quantity of explosive used. True, in the case of Antwerp, it demoralised the civilian population somewhat effectively, which perhaps was the desired end, but the military results were nil.

The Zeppelin, and indeed all dirigibles of large size, have one advantage over aeroplanes. They are able to throw bombs of larger size and charged with greater quantities of high explosive and shrapnel than those which can be hurled from heavier-than-air machines. Thus it has been stated that the largest Zeppelins can drop single charges exceeding one ton in weight, but such a statement is not to be credited.

The shell generally used by the Zeppelin measures about 47 inches in length by 8 1/2 inches in diameter, and varies in weight from 200 to 242 pounds.

Where destruction pure and simple is desired, the shell is charged with a high explosive such as picric acid or T.N.T., the colloquial abbreviation for the devastating agent scientifically known as "Trinitrotoluene," the base of which, in common with all the high explosives used by the different powers and variously known as lyddite, melinite, cheddite, and so forth, is picric acid. Such a bomb, if it strikes the objective, a building, for instance, fairly and squarely, may inflict widespread material damage.

On the other hand, where it is desired to scatter death, as well as destruction, far and wide, an elaborate form of shrapnel shell is utilised. The shell in addition to a bursting charge, contains bullets, pieces of iron, and other metallic fragments. When the shell bursts, their contents, together with the pieces of the shell which is likewise broken up by the explosion, are hurled in all directions over a radius of some 50 yards or more, according to the bursting charge.

These shells are fired upon impact, a detonator exploding the main charge. The detonator, comprising fulminate of mercury, is placed in the head or tail of the missile. To secure perfect detonation and to distribute the death-dealing contents evenly in all directions, it is essential that the bomb should strike the ground almost at right angles: otherwise the contents are hurled irregularly and perhaps in one direction only. One great objection to the percussion system, as the method of impact detonation is called, is that the damage may be localised. A bomb launched from a height of say 1,000 feet attains terrific velocity, due to the force of gravity in conjunction with its own weight, in consonance with the law concerning a falling body, by the time it reaches the ground. It buries itself to a certain depth before bursting so that the forces of the explosion become somewhat muffled as it were. A huge deep hole--a miniature volcano crater--is formed, while all the glass in the immediate vicinity of the explosion may be shattered by the concussion, and the walls of adjacent buildings be bespattered with shrapnel.

Although it is stated that an airship is able to drop a single missile weighing one ton in weight, there has been no attempt to prove the contention by practice. In all probability the heaviest shell launched from a Zeppelin has not exceeded 300 pounds. There is one cogent reason for such a belief. A bomb weighing one ton is equivalent to a similar weight of ballast. If this were discarded suddenly the equilibrium of the dirigible would be seriously disturbed--it would exert a tendency to fly upwards at a rapid speed. It is doubtful whether the planes controlling movement in the vertical plane would ever be able to counteract this enormous vertical thrust. Something would have to submit to the strain. Even if the dirigible displaced say 20 tons, and a bomb weighing one ton were discharged, the weight of the balloon would be decreased suddenly by approximately five per cent, so that it would shoot upwards at an alarming speed, and some seconds would elapse before control was regained.

The method of launching bombs from airships varies considerably. Some are released from a cradle, being tilted into position ready for firing, while others are discharged from a tube somewhat reminiscent of that used for firing torpedoes, with the exception that little or no initial impetus is imparted to the missile; the velocity it attains is essentially gravitational.

The French favour the tube-launching method since thereby it is stated to be possible to take more accurate aim. The objective is sighted and the bomb launched at the critical moment. In some instances the French employ an automatic detonator which corresponds in a certain measure to the time-fuse of a shrapnel shell fired from a gun.

The bomb-thrower reads the altitude of his airship as indicated by his barometer or other recording instrument, and by means of a table at his command ascertains in a moment the time which will elapse before the bomb strikes the ground. The automatic detonator is set in motion and the bomb released to explode approximately at the height to which it is set. When it bursts the full force of the explosion is distributed downwards and laterally. Owing to the difficulty of ensuring the explosion of the bomb at the exact height desired, it is also made to explode upon impact so as to make doubly sure of its efficacy.

Firing timed bombs from aloft, however, is not free from excitement and danger, as the experience of a French airman demonstrates. His dirigible had been commanded to make a night-raid upon a railway station which was a strategical junction for the movement of the enemy's troops. Although the hostile searchlights were active, the airship contrived to slip between the spokes of light without being observed. By descending to a comparatively low altitude the pilot was able to pick up the objective.

Three projectiles were discharged in rapid succession and then the searchlights, being concentrated, struck the airship, revealing its presence to the troops below. Instantly a spirited fusillade broke out. The airmen, by throwing ballast and other portable articles overboard pell-mell, rose rapidly, pursued by the hostile shells.

In the upward travel the bomb-thrower decided to have a parting shot. The airship was steadied momentarily to enable the range to be taken, the automatic detonator was set going and the bomb slipped into the launching tube. But for some reason or other the missile jambed.

The situation was desperate. In a few seconds the bomb would burst and shatter the airship. The bomb-thrower grabbed a tool and climbing into the rigging below hacked away at the bomb- throwing tube until the whole equipment was cut adrift and fell clear of the vessel. Almost instantly there was a terrific explosion in mid-air. The blast of air caused the vessel to roll and pitch in a disconcerting manner, but as the airman permitted the craft to continue its upward course unchecked, she soon steadied herself and was brought under control once more.

The bomb carried by aeroplanes differs consider ably from that used by dirigibles, is smaller and more convenient to handle, though considering its weight and size it is remarkably destructive. In this instance complete reliance is placed upon detonation by impact. The latest types of British war-plane bombs have been made particularly formidable, those employed in the "raids in force" ranging up to 95 pounds in weight.

The type of bomb which has proved to be the most successful is pear-shaped. The tail spindle is given an arrow-head shape, the vanes being utilised to steady the downward flight of the missile. In falling the bomb spins round, the rotating speed increasing as the projectile gathers velocity. The vanes act as a guide, keeping the projectile in as vertical a plane as possible, and ensuring that the rounded head shall strike the ground. The earlier types of bombs were not fitted with these vanes, the result being that sometimes they turned over and over as they fell through the air, while more often than not they failed to explode upon striking the ground.

The method of launching the bomb also varies considerably, experience not having indicated the most efficient method of consummating this end. In some cases the bombs are carried in a cradle placed beneath the aeroplane and launched merely by tilting them in a kind of sling, one by one, to enable them to drop to the ground, this action being controlled by means of a lever. In another instance they are dropped over the side of the car by the pilot, the tail of the bomb being fitted with a swivel and ring to facilitate the operation. Some of the French aviators favour a still simpler method. The bomb is attached to a thread and lowered over the side. At the critical moment it is released simply by severing the thread. Such aeroplane bombs, however, constitute a menace to the machine and to the pilot. Should the bomb be struck by hostile rifle or shell fire while the machine is aloft, an explosion is probable; while should the aero plane make an abrupt descent the missiles are likely to be detonated.

A bomb which circumvents this menace and which in fact will explode only when it strikes the ground is that devised by Mr. Marten-Hale. This projectile follows the usual pear-shape, and has a rotating tail to preserve direction when in flight. The detonator is held away from the main charge by a collar and ball-bearing which are held in place by the projecting end of a screw-releasing spindle. When the bomb is dropped the rotating tail causes the spindle to screw upwards until the projection moves away from the steel balls, thereby allowing them to fall inward when the collar and the detonator are released. In order to bring about this action the bomb must have a fall of at least 200 feet.

When the bomb strikes the ground the detonator falls down on the charge, fires the latter, and thus brings about the bursting of the bomb. The projectile is of the shrapnel type. It weighs 20 pounds complete, is charged with some four pounds of T.N.T., and carries 340 steel balls, which represent a weight of 5 3/4 pounds.

The firing mechanism is extremely sensitive and the bomb will burst upon impact with the hull of an airship, water, or soft soil. This projectile, when discharged, speedily assumes the vertical position, so that there is every probability that it will strike the ground fairly and squarely, although at the same time such an impact is not imperative, because it will explode even if the angle of incidence be only 5 degrees. It is remarkably steady in its flight, the balancing and the design of the tail frustrating completely any tendency to wobble or to turn turtle while falling.

Other types of missile may be used. For instance, incendiary bombs have been thrown with success in certain instances. These bombs are similar in shape to the shrapnel projectile, but are charged with petrol or some other equally highly inflammable mixture, and fitted with a detonator. When they strike the objective the bursting charge breaks up the shell, releasing the contents, and simultaneously ignites the combustible.

Another shell is the smoke-bomb, which, up to the present, has been used only upon a restricted scale. This missile is charged with a certain quantity of explosive to burst the shell, and a substance which, when ignited, emits copious clouds of dense smoke. The scope of such a shell is somewhat restricted, it is used only for the purpose of obstructing hostile artillery fire. The shells are dropped in front of the artillery position and the clouds of smoke which are emitted naturally inter fere with the operations of the gunners. These bombs have also been used with advantage to denote the position of concealed hostile artillery, although their utility in this connection is somewhat uncertain, owing to the difficulty of dropping the bomb so accurately as to enable the range-finders to pick up the range.

Dropping bombs from aloft appears to be a very simple operation, but as a matter of fact it is an extremely difficult matter to strike the target, especially from a high altitude. So far as the aeroplane is concerned it is somewhat at a disadvantage as compared with the airship, as the latter is able to hover over a position, and, if a spring-gun is employed to impart an initial velocity to the missile, there is a greater probability of the projectile striking the target provided it has been well-aimed. But even then other conditions are likely to arise, such as air-currents, which may swing the missile to one side of the objective. Consequently adequate allowance has to be made for windage, which is a very difficult factor to calculate from aloft.

Bomb-dropping from an aeroplane is even more difficult. If for instance the aeroplane is speeding along at 60 miles an hour, the bomb when released will have a speed in the horizontal plane of 60 miles an hour, because momentarily it is travelling at the speed of the aeroplane. Consequently the shell will describe a curved trajectory, somewhat similar to that shown in Fig.

On the other hand, if the aeroplane is travelling slowly, say at 20 miles an hour, the curve of the trajectory will be flatter, and if a head wind be prevailing it may even be swept backwards somewhat after it has lost its forward momentum, and describe a trajectory similar to that in Fig. 8.

A bomb released from an altitude of 1000 feet seldom, if ever, makes a bee-line for the earth, even if dropped from a stationary airship. Accordingly, the airman has to release the bomb before he reaches the target below. The determination of the critical moment for the release is not easy, inasmuch as the airman has to take into his calculations the speed of his machine, his altitude, and the direction and velocity of the air-currents.

The difficulty of aiming has been demonstrated upon several occasions at aviation meetings and other similar gatherings. Monsieur Michelin, who has done so much for aviation in France, offered a prize of L1,00--$5,000--in 1912 for bomb-dropping from an aeroplane. The target was a rectangular space marked out upon the ground, measuring 170 feet long by 40 feet broad, and the missiles had to be dropped from a height of 2,400 feet. The prize was won by the well-known American airman, Lieutenant Riley E. Scott, formerly of the United States Army. He dropped his bombs in groups of three. The first round fell clear of the target, but eight of the remaining missiles fell within the area.

In the German competition which was held at Gotha in September of the same year the results were somewhat disappointing. Two targets were provided. The one represented a military bivouac occupying a superficies of 330 square feet, and the other a captive balloon resembling a Zeppelin. The prizes offered were L500, L200, and L80--$2,500, $1,000 and $400--respectively, and were awarded to those who made the greatest number of hits. The conditions were by no means so onerous as those imposed in the Michelin contest, inasmuch as the altitude limit was set at 660 feet, while no machine was to descend within 165 feet. The first competitor completely failed to hit the balloon. The second competitor flying at 800 feet landed seven bombs within the square, but only one other competitor succeeded in placing one bomb within the space.

Bomb-dropping under the above conditions, however, is vastly dissimilar from such work under the grim realities of war. The airman has to act quickly, take his enemy by surprise, avail himself of any protective covering which may exist, and incur great risks. The opposing forces are overwhelmingly against him. The modern rifle, if fired vertically into the air, will hurl the bullet to a height of about 5,000 feet, while the weapons which have been designed to combat aircraft have a range of 10,000 feet or more.

At the latter altitude aggressive tactics are useless. The airman is unable to obtain a clear sharp view of the country beneath owing to the interference offered to vision by atmospheric haze, even in the dearest of weather. In order to obtain reasonable accuracy of aim the corsair of the sky must fly at about 400 feet. In this respect, however, the aeroplane is at a decided advantage, as compared with the dirigible. The machine offers a considerably smaller target and moves with much greater speed. Experience of the war has shown that to attempt to hurl bombs from an extreme height is merely a waste of ammunition. True, they do a certain amount of damage, but this is due to luck, not judgment.

For success in aerial bomb operations the human element is mainly responsible. The daring airman is likely to achieve the greatest results, as events have proved, especially when his raid is sudden and takes the enemy by surprise. The raids carried out by Marix, Collet, Briggs, Babington, Sippe and many others have established this fact incontrovertibly. In all these operations the airmen succeeded because of their intrepidity and their decision to take advantage of cover, otherwise a prevailing mist or low-lying clouds. Flight-Lieutenant Collet approached the Zeppelin shed at Dusseldorf at an altitude of 6,000 feet. There was a bank of mist below, which he encountered at 1,500 feet. He traversed the depth of this layer and emerged therefrom at a height of only 400 feet above the ground. His objective was barely a quarter of a mile ahead. Travelling at high speed he launched his bombs with what proved to be deadly precision, and disappeared into cover almost before the enemy had grasped his intentions. Lieutenant-Commander, now Flight-Commander, Marix was even more daring. Apparently he had no mist in which to conceal himself but trusted almost entirely to the speed of his machine, which probably at times notched 90 miles per hour. Although his advent was detected and he was greeted with a spirited fusillade he clung to his determined idea. He headed straight for the Zeppelin shed, launched two bombs and swung into the higher reaches of the air without a moment's hesitation. His aim was deadly, since both bombs found their mark, and the Zeppelin docked within was blown up. The intrepid airman experienced several narrow escapes, for his aeroplane was struck twenty times, and one or two of the control wires were cut by passing bullets.

The raid carried out by Commanders Briggs and Babington in company with Lieutenant Sippe upon the Zeppelin workshops at Friedrichshafen was even more daring. Leaving the Allies' lines they ascended to an altitude of 4,500 feet, and at this height held to the pre-arranged course until they encountered a mist, which while protecting them from the alert eyes of the enemy below, was responsible for the separation of the raiders, so that each was forced to act independently and to trust to the compass to bring him out of the ordeal successfully. Lieutenant Sippe sighted Lake Constance, and taking advantage of the mist lying low upon the water, descended to such an extent that he found himself only a few feet above the roofs of the houses. Swinging round to the Lake he descended still lower until at last he was practically skimming the surface of the Lake, since he flew at the amazingly low height of barely seven feet off the water. There is no doubt that the noise of his motor was heard plainly by the enemy, but the mist completely enveloped him, and owing to the strange pranks that fog plays with sound deceived his antagonists.

At last, climbing above the bank of vapour, he found that he had overshot the mark, so he turned quickly and sped backwards. At the same time he discovered that he had been preceded by Commander Briggs, who was bombarding the shed furiously, and who himself was the object of a concentrated fire. Swooping down once more, Lieutenant Sippe turned, rained his bombs upon the objective beneath, drawing fire upon himself, but co-operating with Commander Babington, who had now reached the scene, he manoeuvred above the works and continued the bombardment until their ammunition was expended, when they sped home-wards under the cover of the mist. Considering the intensity of the hostile fire, it is surprising that the aeroplanes were not smashed to fragments. Undoubtedly the high speed of the machines and the zigzagging courses which were followed nonplussed the enemy. Commander Briggs was not so fortunate as his colleagues; a bullet pierced his petrol tank, compelling a hurried descent.

The most amazing feature of these aerial raids has been the remarkably low height at which the airmen have ventured to fly. While such a procedure facilitates marksmanship it increases the hazards. The airmen have to trust implicitly to the fleetness of their craft and to their own nerve. Bearing in mind the vulnerability of the average aeroplane, and the general absence of protective armouring against rifle fire at almost point-blank range, it shows the important part which the human element is compelled to play in bomb-dropping operations.

Another missile which has been introduced by the French airmen, and which is extremely deadly when hurled against dense masses of men, is the steel arrow, or "flechette" as it is called. It is a fiendish projectile consisting in reality of a pencil of solid polished steel, 4 3/4 inches in length. The lower end has a sharp tapering point, 5/8ths of an inch in length. For a distance of 1 1/8th of an inch above this point the cylindrical form of the pencil is preserved, but for the succeeding three inches to the upper end, the pencil is provided with four equally spaced angle flanges or vanes. This flanging of the upper end or tail ensures the arrow spinning rapidly as it falls through the air, and at the same times preserves its vertical position during its descent. The weight of the arrow is two-thirds of an ounce.

The method of launching this fearsome projectile is ingenious. A hundred or even more are packed in a vertical position in a special receptacle, placed upon the floor of the aeroplane, preferably near the foot of the pilot or observer. This receptacle is fitted with a bottom moving in the manner of a trap-door, and is opened by pressing a lever. The aviator has merely to depress this pedal with his foot, when the box is opened and the whole of the contents are released. The fall at first is somewhat erratic, but this is an advantage, as it enables the darts to scatter and to cover a wide area. As the rotary motion of the arrows increases during the fall, the direct line of flight becomes more pronounced until at last they assume a vertical direction free from all wobbling, so that when they alight upon the target they are quite plumb.

When launched from a height they strike the objective with terrific force, and will readily penetrate a soldier's helmet and skull. Indeed, when released at a height of 4,000 feet they have been known to pierce a mounted soldier's head, and pass vertically through his body and that of his horse also. Time after time German soldiers have found themselves pinned to the ground through the arrow striking and penetrating their feet. Owing to the extremely light weight of the darts they can be launched in batches of hundreds at a time, and in a promiscuous manner when the objective is a massed body of infantry or cavalry, or a transport convoy. They are extremely effective when thrown among horses even from a comparatively low altitude, not so much from the fatalities they produce, as from the fact that they precipitate a stampede among the animals, which is generally sufficiently serious and frantic to throw cavalry or a transport-train into wild confusion.

Although aerial craft, when skilfully handled, have proved highly successful as weapons of offence, the possibilities of such aggression as yet are scarcely realised; aerial tactics are in their infancy. Developments are moving rapidly. Great efforts are being centred upon the evolution of more formidable missiles to be launched from the clouds. The airman is destined to inspire far greater awe than at present, to exercise a still more demoralising influence, and to work infinitely more destruction.

CHAPTER XI

ARMOURED AEROPLANES

The stern test of war has served to reveal conclusively the fact that aerial craft can be put out of action readily and effectively, when once the marksman has picked up the range, whether the gunner be conducting his operations with an anti- aircraft gun stationed upon the ground, or from a hostile machine. It will be remembered that Flight-Commander Briggs, on the occasion of the daring British raid upon the Zeppelin sheds at Friedrichshafen, was brought to the ground by a bullet which penetrated his fuel tank. Several other vessels, British, German, French, and Russian alike, have been thrown out of action in a similar manner, and invariably the craft which has been disabled suddenly in this way has fallen precipitately to earth in the fatal headlong dive.

Previous to the outbreak of hostilities there was considerable divergence of opinion upon this subject. The general opinion was that the outspread wings and the stays which constituted the weakest parts of the structure were most susceptible to gun-fire, and thus were likely to fail. But practice has proved that it is the driving mechanism which is the most vulnerable part of the aeroplane.

This vulnerability of the essential feature of the flying machine is a decisive weakness, and exposes the aviator to a constant menace. It may be quite true that less than one bullet in a thousand may hit the machine, but when the lucky missile does find its billet its effect is complete. The fact must not be overlooked that the gunners who work the batteries of anti-aircraft guns are becommg more and more expert as a result of practice, so that as time progresses and improved guns for such duty are rendered available, the work of the aviator is likely to become more dangerous and difficult. Experience has proved that the high velocity gun of to-day is able to hurl its projectile or shell to an extreme height--far greater than was previously considered possible--so that considerable discretion has to be exercised by the airman, who literally bears his life in his hands.

Although elaborate trials were carried out upon the testing ranges with the weapons devised especially for firing upon flying machines, captive balloons being employed as targets, the data thus obtained were neither conclusive nor illuminating. The actual experiences of airmen have given us some very instructive facts upon this point for the first time.

It was formerly held that the zone of fire that is to be considered as a serious danger was within a height of about 4,500 feet. But this estimate was well within the mark. Airmen have found that the modern projectiles devised for this phase of operations are able to inflict distinctly serious damage at an altitude of 9,000 feet. The shell itself may have but little of its imparted velocity remaining at this altitude, but it must be remembered that when the missile bursts, the contents thereof are given an independent velocity, and a wide cone of dispersion, which is quite sufficient to achieve the desired end, inasmuch as the mechanism of the modern aeroplane and dirigible is somewhat delicate.

It was for this reason that the possibility of armouring the airship was discussed seriously, and many interesting experiments in this field were carried out. At the same time it was decided that the armouring should be effected upon lines analogous to that prevailing in warship engineering. The craft should not only be provided with defensive but also with aggressive armament. This decision was not viewed with general approbation. It was pointed out that questions of weight would arise, especially in relation to the speed of the machine. Increased weight, unless it were accompanied by a proportionate augmentation of power in the motor, would react against the efficiency and utility of the machine, would appreciably reduce its speed, and would affect its climbing powers very adversely. In some quarters it was maintained that as a result the machine would even prove unsuited to military operations, inasmuch as high speed is the primary factor in these.

Consequently it was decided by the foremost aviating experts that machines would have to be classified and allotted to particular spheres of work, just as warships are built in accordance with the special duty which they are expected to perform. In reconnaissance, speed is imperative, because such work in the air coincides with that of the torpedo-boat or scout upon the seas. It is designed to acquire information respecting the movements of the enemy, so as to assist the heavier arms in the plan of campaign. On the other hand, the fighting corsair of the skies might be likened to the cruiser or battleship. It need not possess such a high turn of speed, but must be equipped with hard-hitting powers and be protected against attacking fire.

One attempt to secure the adequate protection against gun-fire from the ground assumed the installation of bullet-proof steel plating, about one fifth of an inch thick, below the tank and the motor respectively. The disposition of the plating was such as to offer the minimum of resistance to the air and yet to present a plane surface to the ground below. So far as it went this protection was completely effective, but it failed to armour the vital parts against lateral, cross and downward fire while aloft. As the latter is more to be feared than the fire from the ground, seeing that it may be directed at point blank range, this was a decided defect and the armour was subsequently abandoned as useless.

The only effective method of achieving the desired end is to armour the whole of the carriage or fuselage of the adroplane, and this was the principle adopted by the Vickers Company. The Vickers military aeroplane is essentially a military machine. It is built of steel throughout. The skeleton of the machine is formed of an alloy which combines the qualities of aluminium and steel to ensure toughness, strength, and lightness. In fact, metal is employed liberally throughout, except in connection with the wings, which follow the usual lines of construction. The body of the car is sheathed with steel plating which is bullet proof against rifle or even shrapnel fire. The car is designed to carry two persons; the seats are therefore disposed tandemwise, with the observer or gunner occupying the front seat.

The defensive armament is adequate for ordinary purposes. Being fitted with a 100 horse-power motor, fairly high speeds are attainable, although the velocity is not equal to that of machines constructed upon conventional lines, inasmuch as there is an appreciable increase in weight.

The car is short and designed upon excellent stream lines, so that the minimum of resistance to the air is offered, while at the same time the balancing is perfect. The sides of the car are brought up high enough to protect the aviators, only their heads being visible when they are seated. The prow of the car follows the lines generally adopted in high speed torpedo boat design; there is a sharp knife edge stem with an enclosed fo'c's'le, the latter housing the gun.

Another craft, designed for scouting operations, may be likened to the mosquito craft of the seas. This machine, while a biplane like the military aeroplane, is of lighter construction, everything being sacrificed to speed in this instance. It is fitted with a 100 horse-power motor and is designed to carry an observer if required. There is no offensive armament, however. The fuel tank capacity, moreover, is limited, being only sufficient for a two or three hours' flight. While this is adequate for general reconnoitring, which for the most part entails short high speed flights, there are occasions when the Staff demands more prolonged observations conducted over a greater radius. This requisition can be met by eliminating the observer, whose duties in this instance must be assumed by the pilot, and substituting in place of the former, a second fuel tank of sufficient capacity for a flight of four or five hours, thereby bringing the term of action in the air to about 6 1/4 hours. This machine travels at a very high speed and is eminently adapted to its specific duty, but it is of limited service for general purposes.

The arming of an aeroplane, to enable it to defend itself against hostile attack or to participate in raiding operations upon the aerial fleet of the enemy, appears to be a simple task, but as a matter of fact it is an undertaking beset with difficulties innumerable. This is especially the case where the aeroplane is of the tractive type, that is to say where the propellers are placed in the forefront of the machine and in their revolution serve to draw the machine forward. All other considerations must necessarily be sacrificed to the mounting of the propeller. Consequently it is by no means easy to allot a position for the installation of a gun, or if such should be found there is grave risk of the angle of fire being severely restricted. In fact, in many instances the mounting of a gun is out of the question: it becomes a greater menace to the machine than to the enemy.

The French aeronautical section of the military department devoted considerable study to this subject, but found the problem almost insurmountable. Monsieur Loiseau met with the greatest measure of success, and his system is being practised in the present campaign. This principle is essentially adapted to tractor aeroplanes. Forward of the pilot a special position is reserved for the gunner. A special mounting is provided towards the prow, and upon the upper face of the body of the machine. The gun mounting is disposed in such a manner that it is able to command a wide arc of fire in the vertical plane over the nose of the machine and more particularly in the downward direction.

The marksman is provided with a special seat, but when he comes into action he has to stand to manipulate his weapon. The lower part of his body is protected by a front shield of steel plate, a fifth of an inch in thickness, while a light railing extending upon either side and behind enables the gunner to maintain his position when the aeroplane is banking and climbing. The machine gun, of the Hotchkiss type, is mounted upon a swivel attached to a tripod, while the latter is built into the bracing of the car, so as to ensure a fairly steady gun platform.

While the gun in the hands of a trained marksman may be manipulated with destructive effect, the drawbacks to the arrangement are obvious. The gunner occupies a very exposed position, and, although the bullet-proof shield serves to break the effects of wind when travelling at high speed which renders the sighting and training of the weapon extremely difficult, yet he offers a conspicuous target, more particularly when the enemy is able to assume the upper position in the air as a result of superior speed in travelling. The gun, however, may be elevated to about 60 degrees, which elevation may be accentuated by the inclination of the aeroplane when climbing, while the facility with which the weapon may be moved through the horizontal plane is distinctly favourable.

But the aerial marksman suffers from one very pronounced defect: he has a severely restricted survey of everything below, since his vision is interrupted by the planes. The result is that an enemy who has lost ascendancy of position is comparatively safe if he is able to fly immediately below his adversary: the mitrailleuse of the latter cannot be trained upon him. On the other hand the enemy, if equipped with repeating rifles or automatic pistols, is able to inflict appreciable damage upon the craft overhead, the difficulties of firing vertically into the air notwithstanding.

In the Vickers system, where the propeller is mounted behind the car, the aeroplane thus operating upon the pusher principle, the nose of the car is occupied by the arm, which is a rifle calibre machine gun fitted upon a special mounting. The prow is provided with an embrasure for the weapon and the latter is so installed as to command an angle of 30 degrees on all sides of the longitudinal axis of the machine when in flight. In this instance the marksman is provided with complete protection on all sides, inasmuch as his position is in the prow, where the hood of the fo'c's'le shields him from overhead attack. The gun is protected by a special shield which moves with the gun barrel. This shield is provided with mica windows, through which the gunner is able to sight his arm, so that he is not inconvenienced in any way by the wind draught.

One shortcoming of such methods of arming an aeroplane will be observed. Ahead firing only is possible; the weapon cannot be trained astern, while similarly the line of fire on either broadside is severely limited. This is one reason why the machine-gun armament of aerial craft of the heavier-than-air type has not undergone extensive development. In many instances the pilot and observer have expressed their preference for repeating high velocity rifles over any form of fixed gun mounting, and have recourse to the latter only when the conditions are extremely favourable to its effective employment.

Efforts are now being made to equip the military type of aeroplane with both forward and astern firing guns. The urgency of astern fire has been brought home very vividly. Suppose, for instance, two hostile aeroplanes, A and B, are in the air. A has the advantage at first, but B is speedier and rapidly overhauls A. During the whole period of the overhauling movement the gun of B can be directed upon A, while the latter, owing to the arc of training being limited to c d cannot reply. Obviously in the running fight it would be to the advantage of B, although the fleeter machine, to keep behind A (position 1), but the latter is making towards its own lines.

Under these circumstances A must be headed off, so B crowds on speed to consummate this end. But in the overtaking process B renders his gun-fire ineffective, inasmuch as B passes beyond the arc of his gun which is represented by e f. But in so doing B comes within the firing arc of A (position 9). To minimise this danger B ascends to a higher level to obtain the paramount position.

If, however, B were equipped with an astern gun the aeroplane A would be within the fire of B when the forward gun of the latter could not be used. Similarly if A were also fitted with an astern gun it would be able to attack its pursuer the whole time B was to its rear and in this event, if its gun-fire were superior, it would be able to keep the latter to a safe distance, or compel B to manoeuvre into a superior position, which would entail a certain loss of time.

An astern firing gun would be valuable to B in another sense. Directly it had passed A or brought the latter within the zone of its astern gun it could maintain its fire at the most advantageous range, because owing to its speed it would be able to dictate the distance over which shots should be exchanged and if mounted with a superior weapon would be able to keep beyond the range of A's guns while at the same time it would keep A within range of its own gun and consequently rake the latter. In the interests of self-preservation A would be compelled to change its course; in fact, B would be able to drive it in any direction he desired, as he would command A's movements by gun-fire.

The value of combined ahead and astern firing has been appreciated, but there is one difficulty which at the moment appears to be insuperable the clearance of the propeller. At the moment astern-firing, if such it may be called, is maintained by repeating rifles, but this armament is not to be compared with machine-gun firing, as the latter with its capacity to pour 400 to 600 shots a minute, is far more deadly, particularly when the weapon is manipulated by a crack gunner.

Up to the present the offensive armament of aeroplanes has been confined to light machine guns such as the Hotchkiss, Berthier, Schwartlose, and Maxim weapons. So far as the arming of aeroplanes is concerned the indispensable condition is light weight. With airships this factor is not so vital, the result being that some dirigibles are mounted with guns, throwing one pound bursting shells, fitted either with delay action or percussion fuses, the former for preference. These shells are given a wide cone of dispersion. Experiments are also being made with a gun similar to the pom-pom which proved so useful in South Africa, the gun throwing small shells varying from four to eight ounces in weight at high velocity and in rapid succession. While such missiles would not be likely to inflict appreciable damage upon an armoured aeroplane, they would nevertheless be disconcerting to the aviators subjected to such fire, and in aerial combats the successful undermining of the adversary's moral is of far greater importance than in land operations, since immediately ascendancy in the artillery operations is attained the final issue is a matter of moments.

But the most devastating arm which has yet been contrived for aerial operations is the light machine gun which has recently been perfected. The one objective with this weapon is to disable the hostile aircraft's machinery. It fires an armour piercing projectile which, striking the motor of any aircraft, would instantly put the latter out of action. The shell has a diameter of about .75 inch and weighs about four ounces. The gun is a hybrid of the mitrailleuse and the French "Soixante-quinze," combining the firing rapidity of the former with the recoil mechanism of the latter. This missile has established its ability to penetrate the defensive armouring of any aeroplane and the motor of the machine at 1,000 yards' range. This offensive arm is now being manufactured, so that it is likely to be seen in the near future as the main armament of aeroplanes.

At the moment widespread efforts are being made in the direction of increasing the offensive efficiency of aircraft. It is one of the phases of ingenuity which has been stimulated into activity as a result of the war.

CHAPTER XII

## BATTLES IN THE AIR

Ever since the days of Jules Verne no theme has proved so popular in fiction as fighting in the air. It was a subject which lent itself to vivid imagination and spirited picturesque portrayal. Discussion might be provoked, but it inevitably proved abortive, inasmuch as there was a complete absence of data based upon actual experience. The novelist was without any theory: he avowedly depended upon the brilliance of his imagination. The critic could only theorise, and no matter how dogmatic his reasonings, they were certainly as unconvincing as those of the object of his attack.

But truth has proved stranger than fiction. The imaginative pictures of the novelist have not only been fulfilled but surpassed, while the theorising critic has been utterly confounded. Fighting in the air has become so inseparable from the military operations of to-day that it occurs with startling frequency. A contest between hostile aeroplanes, hundreds of feet above the earth, is no longer regarded as a dramatic, thrilling spectacle: it has become as matter-of-fact as a bayonet melee between opposed forces of infantry.

A duel in the clouds differs from any other form of encounter. It is fought mercilessly: there can be no question of quarter or surrender. The white flag is no protection, for the simple reason that science and mechanical ingenuity have failed, so far, to devise a means of taking an aeroplane in tow. The victor has no possible method of forcing the vanquished to the ground in his own territory except driving. If such a move be made there is the risk that the latter will take the advantage of a critical opportunity to effect his escape, or to turn the tables. For these reasons the fight is fought to a conclusive finish.

To aspire to success in these combats waged in the trackless blue, speed, initiative, and daring are essential. Success falls to the swift in every instance. An aeroplane travelling at a high speed, and pursuing an undulating or irregular trajectory is almostimpossible to hit from the ground, as sighting is so extremely difficult. Sighting from another machine, which likewise is travelling rapidly, and pursuing an irregular path, is far more so. Unless the attacker can approach relatively closely to his enemy the possibility of hitting him is extremely remote. Rifle or gun-fire must be absolutely point blank.

When a marauding aeroplane is espied the attacking corsair immediately struggles for the strategical position, which is above his adversary. To fire upwards from one aeroplane at another is virtually impossible, at least with any degree of accuracy. The marksman is at a hopeless disadvantage. If the pilot be unaccompanied and entirely dependent upon his own resources he cannot hope to fire vertically above him, for the simple reason that in so doing he must relinquish control of his machine. A rifle cannot possibly be sighted under such conditions, inasmuch as it demands that the rifleman shall lean back so as to obtain control of his weapon and to bring it to bear upon his objective. Even if a long range Mauser or other automatic pistol of the latest type be employed, two hands are necessary for firing purposes, more particularly as, under such conditions, the machine, if not kept under control, is apt to lurch and pitch disconcertingly.

Even a colleague carried for the express purpose of aggression is handicapped. If he has a machinegun, such as a Maxim or a mitrailleuse, it is almost out of the question to train it vertically. Its useful vertical training arc is probably limited to about 80 degrees, and at this elevation the gunner has to assume an extremely uncomfortable position, especiauy upon an aeroplane, where, under the best of circumstances, he is somewhat cramped.

On the other hand the man in the aeroplane above holds the dominating position. He is immediately above his adversary and firing may be carried out with facility. The conditions are wholly in his favour. Sighting and firing downwards, even if absolutely vertically, imposes the minimum physical effort, with the result that the marksman is able to bring a steadier aim upon his adversary. Even if the machine be carrying only the pilot, the latter is able to fire upon his enemy without necessarily releasing control of his motor, even for a moment.

If he is a skilled sharpshooter, and the exigencies demand, he can level, sight, and fire his weapon with one hand, while under such circumstances an automatic self-loading pistol can be trained upon the objective with the greatest ease. If the warplane be carrying a second person, acting as a gunner, the latter can maintain an effective rifle fusillade, and, at the same time, manipulate his machine-gun with no great effort, maintaining rifle fire until the pilot, by manoeuvring, can enable the mitrailleuse or Maxim to be used to the greatest advantage.

Hence the wonderful display of tactical operations when two hostile aeroplanes sight one another. The hunted at first endeavours to learn the turn of speed which his antagonist commands. If the latter is inferior, the pursued can either profit from his advantage and race away to safety, or at once begin to manoeuvre for position. If he is made of stern stuff, he attempts the latter feat without delay. The pursuer, if he realises that he is out classed in pace, divines that his quarry will start climbing if he intends to show fight, so he begins to climb also.

Now success in this tactical move will accrue to the machine which possesses the finest climbing powers, and here again, of course, speed is certain to count. But, on the other hand, the prowess of the aviator--the human element once more--must not be ignored. The war has demonstrated very convincingly that the personal quality of the aviator often becomes the decisive factor.

A spirited contest in the air is one of the grimmest and most thrilling spectacles possible to conceive, and it displays the skill of the aviator in a striking manner. Daring sweeps, startling wheels, breathless vol-planes, and remarkable climbs are carried out. One wonders how the machine can possibly withstand the racking strains to which it is subjected. The average aeroplane demands space in which to describe a turn, and the wheel has to be manipulated carefully and dexterously, an operation requiring considerable judgment on the part of the helmsman.

But in an aerial duel discretion is flung to the winds. The pilot jambs his helm over in his keen struggle to gain the superior position, causing the machine to groan and almost to heel over. The stem stresses of war have served to reveal the perfection of the modern aeroplane together with the remarkable strength of its construction. In one or two instances, when a victor has come to earth, subsequent examination has revealed the enormous strains to which the aeroplane has been subjected. The machine has been distorted; wires have been broken--wires which have succumbed to the enormous stresses which have been imposed and have not been snapped by rifle fire. One well-known British airman, who was formerly a daring automobilist, confided to me that a fight in the air "is the finest reliability trial for an aeroplane that was ever devised!"

In these desperate struggles for aerial supremacy the one party endeavours to bring his opponent well within the point-blank range of his armament: the other on his part strives just as valiantly to keep well out of reach. The latter knows fully well that his opponent is at a serious disadvantage when beyond point-blank range, for the simple reason that in sighting the rifle or automatic pistol, it is difficult, if not impossible while aloft, to judge distances accurately, and to make the correct allowances for windage.

If, however, the dominating aviator is armed with a machine gun he occupies the superior position, because he can pour a steady hail of lead upon his enemy. The employment of such a weapon when the contest is being waged over friendly territory has many drawbacks. Damage is likely to be inflicted among innocent observers on the earth below; the airman is likely to bombard his friends. For this very reason promiscuous firing, in the hope of a lucky shot finding a billet in the hostile machine, is not practised. Both parties appear to reserve their fire until they have drawn within what may be described as fighting distance, otherwise point blank range, which may be anything up to 300 yards.

Some of the battles between the German and the French or British aeroplanes have been waged with a total disregard of the consequences. Both realise that one or the other must perish, and each is equally determined to triumph. It is doubtful whether the animosity between the opposing forces is manifested anywhere so acutely as in the air. In some instances the combat has commenced at 300 feet or so above the earth, and has been fought so desperately, the machines climbing and endeavouring to outmanoeuvre each other, that an altitude of over 5,000 feet has been attained before they have come to close grips.

The French aviator is nimble, and impetuous: the German aviator is daring, but slow in thought: the British airman is a master of strategy, quick in thought, and prepared to risk anything to achieve his end. The German airman is sent aloft to reconnoitre the enemy and to communicate his information to his headquarters. That is his assigned duty and he performs it mechanically, declining to fight, as the welfare of his colleagues below is considered to be of more vital importance than his personal superiority in an aerial contest. But if he is cornered he fights with a terrible and fatalistic desperation.

The bravery of the German airmen is appreciated by the Allies. The French flying-man, with his traditional love for individual combat, seeks and keenly enjoys a duel. The British airman regards such a contest as a mere incident in the round of duty, but willingly accepts the challenge when it is offered. It is this manifestation of what may be described as acquiescence in any development that enabled the British flying corps, although numerically inferior, to gain its mastery of the air so unostentatiously and yet so completely.

All things considered an aeroplane duel is regarded as a fairly equal combat. But what of a duel between an aeroplane and a dirigible? Which holds the advantage? This question has not been settled, at any rate conclusively, but it is generally conceded that up to a certain point the dirigible is superior. It certainly offers a huge and attractive target, but rifle fire at its prominent gas-bag is not going to cause much havoc. The punctures of the envelope may represent so many vents through which the gas within may effect a gradual escape, but considerable time must elapse before the effect of such a bombardment becomes pronounced in its result, unless the gas-bag is absolutely riddled with machine gun-fire, when descent must be accelerated.

On the other hand, it is to be presumed that the dirigible is armed. In this event it has a distinct advantage. It has a steady gun-platform enabling the weapons of offence to be trained more easily and an enhanced accuracy of,fire to be obtained. In order to achieve success it is practically imperative that an aeroplane should obtain a position above the dirigible, but the latter can ascend in a much shorter space of time, because its ascent is vertical, whereas the aeroplane must describe a spiral in climbing. Under these circumstances it is relatively easy for the airship to outmanoeuvre the aeroplane in the vertical plane, and to hold the dominating position.

But even should the aeroplane obtain the upper position it is not regarded with fear. Some of the latest Zeppelins have a machine gun mounted upon the upper surface of the envelope, which can be trained through 360 degrees and elevated to about 80 degrees vertical. Owing to the steady gun platform offered it holds command in gun-fire, so that the aeroplane, unless the aviator is exceptionally daring, will not venture within the range of the dirigible. It is stated, however, that this upper gun has proved unsatisfactory, owing to the stresses and strains imposed upon the framework of the envelope of the Zeppelin during firing, and it has apparently been abandoned. The position, however, is still available for a sniper or sharpshooter.

The position in the sky between two such combatants is closely analogous to that of a torpedo boat and a Dreadnought. The latter, so long as it can keep the former at arm's, or rather gun's, distance is perfectly safe. The torpedo boat can only aspire to harass its enemy by buzzing around, hoping that a lucky opportunity will develop to enable it to rush in and to launch its torpedo. It is the same with the aeroplane when arrayed against a Zeppelin. It is the mosquito craft of the air.

How then can a heavier-than-air machine triumph over the unwieldy lighter-than-air antagonist? Two solutions are available. If it can get above the dirigible the adroplane may bring about the dirigible's destruction by the successful launch of a bomb. The detonation of the latter would fire the hydrogen within the gas-bag or bags, in which event the airship would fall to earth a tangled wreck. Even if the airship were inflated with a non-inflammable gas--the Germans claim that their Zeppelins now are so inflated--the damage wrought by the bomb would be so severe as to destroy the airship's buoyancy, and it would be forced to the ground.

The alternative is very much more desperate. It involves ramming the dirigible. This is undoubtedly possible owing to the speed and facile control of the aeroplane, but whether the operation would be successful remains to be proved. The aeroplane would be faced with such a concentrated hostile fire as to menace its own existence--its forward rush would be frustrated by the dirigible just as a naval vessel parries the ramming tactics of an enemy by sinking the latter before she reaches her target, while if it did crash into the hull of the dirigible, tearing it to shreds, firing its gas, or destroying its equilibrium, both protagonists would perish in the fatal dive to earth. For this reason ramming in mid-air is not likely to be essayed except when the situation is desperate.

What happens when two aeroplanes meet in dire combat in mid-air and one is vanquished? Does the unfortunate vessel drop to earth like a stone, or does it descend steadily and reach the ground uninjured? So far as actual experience has proved, either one of the foregoing contingencies may happen. In one such duel the German aeroplane was observed to start suddenly upon a vol-plane to the ground. Its descending flight carried it beyond the lines of the Allies into the territory of its friends. Both came to the conclusion that the aviator had effected his escape. But subsequent investigation revealed the fact that a lucky bullet from the Allies' aeroplane had lodged in the brain of the German pilot, killing him instantly. At the moment when Death over took him the aviator had set his plane for the descent to the ground, and the machine came to earth in the manner of a glider.

But in other instances the descent has been far more tragic. The aeroplane, deprived of its motive power, has taken the deadly headlong dive to earth. It has struck the ground with terrific violence, burying its nose in the soil, showing incidentally that a flying machine is an indifferent plough, and has shattered itself, the debris soaked with the escaping fuel becoming ignited. In any event, after such a fall the machine is certain to be a wreck. The motor may escape damage, in which event it is salvaged, the machine subsequently being purposely sacrificed to the flames, thereby rendering it no longer available to the enemy even if captured. In many instances the hostile fire has smashed some of the stays and wires, causing the aeroplane to lose its equilibrium, and sending it to earth in the manner of the proverbial stone, the aviators either being dashed to pieces or burned to death.

What are the vulnerable parts of the aeroplane? While the deliberate intention of either combatant is to put his antagonist hors de combat, the disablement of the machine may be achieved without necessarily killing or even seriously wounding the hostile airman. The prevailing type of aeroplane is highly susceptible to derangement: it is like a ship without armour plate protection. The objective of the antagonist is the motor or the fuel-tank, the vital parts of the machine, as much as the aviator seated within.

A well-planted shot, which upsets the mechanism of the engine, or a missile which perforates the fuel tank, thereby depriving the motor of its sustenance, will ensure victory as conclusively as the death of the aviator himself. Rifle fire can achieve either of these ends with little difficulty. Apart from these two nerve-centres, bombardment is not likely to effect the desired disablement, inasmuch as it cannot be rendered completely effective. The wings may be riddled like a sieve, but the equilibrium of the machine is not seriously imperilled thereby. Even many of the stays may be shot away, but bearing in mind the slender objective they offer, their destruction is likely to be due more to luck than judgment. On the other hand, the motor and fuel tank of the conventional machine offer attractive targets: both may be put out of action readily, and the disablement of the motive power of an enemy's craft, be it torpedo-boat, battleship, or aeroplane, immediately places the same at the assailant's mercy.

Nevertheless, of course, the disablement of the airman brings about the desired end very effectively. It deprives the driving force of its controlling hand; The aeroplane becomes like a ship without a rudder: a vessel whose helmsman has been shot down. It is unmanageable, and likely to become the sport of the element in which it moves. It is for this reason that aviators have been urged to direct their fire upon the men and mechanism of a dirigible in the effort to put it out of action. An uncontrolled airship is more likely to meet with its doom than an aeroplane. The latter will inevitably glide to earth, possibly damaging itself seriously in the process, as events in the war have demonstrated, but a helpless airship at once becomes the sport of the wind, and anyone who has assisted, like myself, in the descent of a vessel charged with gas and floating in the air, can appreciate the difficulties experienced in landing. An uncontrolled Zeppelin, for instance, would inevitably pile up in a tangled twisted ruin if forced to descend in the manner of an ordinary balloon. Consequently the pilot of a dirigible realises to the full the imperative urgency of keeping beyond the point-blank fire of aerial mosquito craft.

The assiduity with which British aviators are prepared to swarm to the attack has been responsible for a display of commendable ingenuity on the part of the German airman. Nature has provided some of its creatures, such as the octopus, for instance, with the ways and means of baffling its pursuers. It emits dense clouds of inky fluid when disturbed, and is able to effect its escape under cover of this screen.

The German aviator has emulated the octopus. He carries not only explosive bombs but smoke balls as well. When he is pursued and he finds himself in danger of being overtaken, the Teuton aviator ignites these missiles and throws them overboard. The aeroplane becomes enveloped in a cloud of thick impenetrable smoke. It is useless to fire haphazard at the cloud, inasmuch as it does not necessarily cover the aviator. He probably has dashed out of the cloud in such a way as to put the screen between himself and his pursuer.

In such tactics he has merely profited by a method which is practised freely upon the water. The torpedo boat flotilla when in danger of being overwhelmed by superior forces will throw off copious clouds of smoke. Under this cover it is able to steal away, trusting to the speed of the craft to carry them well beyond gunshot. The "smoke screen," as it is called, is an accepted and extensively practised ruse in naval strategy, and is now adopted by its mosquito colleagues of the air.

CHAPTER XIII

TRICKS AND RUSES TO BAFFLE THE AIRMAN

The airman has not been allowed to hold his undisputed sway in military operations for long. Desperate situations demand drastic remedies and already considerable and illuminating ingenuity is being displayed to baffle and mislead the scout of the skies.

It is a somewhat curious and noteworthy fact, that the Germans were among the first to realise the scope of the airman's activities, and the significance of their relation to the conveyance of intimate information and the direction of artillery fire. Consequently, they now spare no effort to convey illusory information, in the hope that the hostile force may ultimately make a false move which may culminate in disaster.

Thus, for instance, as much endeavour is bestowed upon the fashioning of dummy trenches as upon the preparation of the actual lines of defence. And every care will be taken to indicate that the former are strongly held. The dug-outs are complete and at places are apparently cunningly masked. If the airman is flying swiftly, he is likely to fail to distinguish the dummy from the real trenches. To him the defences appear to be far more elaborate and more strongly held than is the actual case.

The advantage of this delusion is obvious when a retreat is being made. It enables the enemy to withdraw his forces deliberately and in perfect order, and to assume another and stronger position comparatively at leisure. The difficulty of detecting the dummies is emphasised, inasmuch as now, whenever the sound of an aeroplane is heard, or a glimpse thereof is obtained, the men keep well down and out of sight. Not a sign of movement is observable. For all the airman may know to the contrary, the trenches may be completely empty, whereas, as a matter of fact, they are throbbing with alert infantry, anxious for a struggle with the enemy.

This is one instance where the dirigible is superior to the aeroplane. The latter can only keep circling round and round over the suspicious position; the movement through the air interferes with close continuous observation. On the other hand, the dirigible can maintain a stationary position aloft for hours on end. Then the issue is resolved into a contest of patience, with the advantage to the airman. The soldiers in the trenches fret and fume under cover; confined concealment is irksome and is a supreme test of the nerves. Unless the soldiers are made of very stern stuff, physical endurance succumbs. Some rash act-- apparently very trivial--may be committed; it suffices for the vigilant sentinel overhead. He detects the slender sign of life, forms his own conclusions, and returns to his headquarters with the intelligence that the enemy is playing "Brer Rabbit."

It has also become increasingly difficult for the airman to gather absolutely trustworthy data concerning the disposition and movement of troops. Small columns are now strung out along the highways to convey the impression that the moving troops are in far greater force than is actually the case, while the main body is under the cover offered by a friendly wood and is safe from detection. The rapidity with which thousands of men are able to disappear when the word "Airman" is passed round is astonishing. They vanish as completely and suddenly as if swallowed by the earth or dissolved into thin air. They conceal themselves under bushes,in ditches, lie prone under hedgerows, dart into houses and outbuildings--in short, take every cover which is available, no matter how slender it may seem, with baffling alacrity. The attenuated column, however, is kept moving along the highway for the express purpose of deceiving the airman.

Advancing troops also are now urged to move forward under the shelter of trees, even if the task entails marching in single or double file, to escape the prying eyes of the man above. By keeping close to the line of trunks, thus taking full advantage of the overhanging branches, and marching in such a manner as to create little dust, it is possible to escape the aerial scout.

The concealment of cavalry, however, is somewhat difficult. An animal, especially if he be unaccustomed to the noise of the aeroplane, is likely to become startled, and to give vent to a frightened and vociferous neighing which invariably provokes a hearty response from his equine comrades. The sharp ear of the airman does not fail to distinguish this sound above the music of his motor. Again, he has come to regard all copses and stretches of undergrowth with suspicion. Such may or may not harbour the enemy, but there is no risk in making an investigation. He swoops down, and when a short distance above the apparently innocent copse, circles round it two or three times. Still undecided, he finally hurls a bomb. Its detonation invariably proves effective. The horses stampede and the secret is out. Even foot soldiers must be severely trained and experienced to resist the natural inclination to break cover when such a missile is hurled into their midst.

Frequently a force, which has laboured under the impression that it is safe from detection, has revealed its presence unwittingly and upon the spur of the moment. If the men be steeled against the bomb attack, it is almost impossible to resist the inclination to take a shot when the airman, swooping down, ventures so temptingly near as to render him an enticing target almost impossible to miss. As a rule, however, the observer is on the alert for such a betrayal of a force's existence. When the bomb fails to scatter the enemy, or the men are proof against the temptation to fire a volley, a few rounds from the aeroplane's machine gun often proves effective. If the copse indeed be empty no harm is done, beyond the abortive expenditure of a few rounds of ammunition: if it be occupied, the fruits of the manoeuvre are attractive. Cunning is matched against cunning, and the struggle for supremacy in the art of craftiness is keen.

The French Flying Corps have had recourse to an ingenious ruse for accomplishing two ends--the one to draw concealed artillery fire, and the other to pre-occupy the airmen. Two German aerial scouts observed a French machine flying at a somewhat venturesome height over their masked artillery. Divining the reason for the hostile intrepidity they gave chase. Circling round the French machine they assailed it with machine-gun fire. The enemy appeared to take no notice but continued his gradual descent in a steady line.

Presently the German airmen, having drawn sufficiently near, observed that the French aviator was inert. Had he been killed? Everything pointed to such a conclusion, especially as they had raked the aeroplane fore and aft with bullets. But still suspicious they continued their circling movements, their attention so concentrated upon their quarry that they had not observed another move. It was the crash of guns from their masked artillery which broke in upon their absorption. Looking round, they observed three French aeroplanes soaring around and above them at high speed. Scarcely had they realised the situation before a spirited mitraireuse fire was rained upon them. One of the German aeroplanes was speedily disabled. Its fuel tank was riddled and it sank rapidly, finally crashing to earth in the deadly dive head foremost, and killing both its occupants in the fall. The second aeroplane hurried away with its pilot wounded. In the excitement of the aerial melee the first French aeroplane had been forgotten. It was now scarcely 100 feet above the German artillery. A capture appeared to be imminent, but the Germans received a rude surprise. Suddenly the aeroplane exploded and a hail of shrapnel burst over the heads of the artillerymen.

The circumstance was decidedly uncanny, but after two or three such experiences of exploding aeroplanes the matter was explained. The apparently helpless aeroplane was merely a glider, which, instead of carrying a man, had a booby-trap aboard.

It appears that the French airmen have found a use for the aeroplanes which are considered unsafe for further use. The motor and propeller are removed and the dummy of explosives is strapped into position. The laden glider is then taken aloft by means of an airship, and in the concealment of the clouds is released, the rudder being so set as to ensure a gradual vol-plane towards the suspicious position below. The explosive cargo is set with a time fuse, the arrangement being that the contents will be detonated while the machine is near the ground, unless this end is accelerated by a well-planted shell from an anti-aircraft gun. The decoy glider is generally accompanied by one or two aeroplanes under control, which keep under the cover of the clouds until the hostile aviators have been drawn into the air, when they swoop down to the attack. The raiders are fully aware that they are not likely to become the target of fire from the ground, owing to the fact that the enemy's artillery might hit its friends. Consequently the antagonistic airmen are left to settle their own account. In the meantime the dummy machine draws nearer to the ground to explode and to scatter its death-dealing fragments of steel, iron, and bullets in all directions.

Possibly in no other phase of warfare is subterfuge practised so extensively as in the concealment of guns. The branches of trees constitute the most complete protection and guns are placed in position beneath a liberal cover of this character. The branches also offer a screen for the artillerymen, who can lurk beneath this shelter until the aeroplane has passed. To complete the illusion dummy guns fashioned from tree trunks and the wheels of useless limbers are rigged up, and partially hidden under branches, the idea being to convey the impression to the man aloft that they are the actual artillery.

The aerial scout observes the dummies beneath the sparse covering of branches. Congratulating himself upon his sharp eyesight, he returns to his base with the intelligence that he has found the enemy's guns he indicates their position upon the map, and in some cases returns to notify the position of the weapons by smoke-ball or tinsel, when they are immediately subjected to a severe bombardment. He follows the shell-fire and sees the arms put out of action. He returns to camp satisfied with his exploit, oblivious of the smiles and laughter of the hostile artillerymen, who have their guns safely in position and well masked some distance away. The dummies are imperfectly concealed purposely, so that they may be discovered by the aerial scout, while the real guns are completely masked and ready to belch forth from another point. In one or two cases the dummies have been rigged up in such a manner as to convey the impression, when seen from aloft, that a whole battery has been put out of action, barrels and wheels as well as broken limbers strewing the ground in all directions.

Moving masses of soldiers are also resorting to cunning in order to mislead the airman or to escape his observation. At the battle of Haelen, during which engagement the German warplanes were exceptionally active, the Belgian soldiers covered their heads with bundles of wheat snatched from the standing stooks, and under this cover lurked in a field where the corn was still standing. From aloft their forms defied detection: the improvised headgear completely covered them and blended effectively with the surrounding wheat. In another instance the French misled a German airman somewhat effectively. What appeared to be cavalry was seen to be retreating along the country road, and the airman returned hurriedly to report. A German squadron was dispatched in hasty pursuit. But as it rounded a copse skirting the road it received a murderous fire at close quarters, which decimated the ranks and sent the survivors flying for their lives along the road up which they had ridden so confidently. Had the aviator been in a position to observe the horses more closely, he would have found that what appeared to be riders on their backs were in reality sacks stuffed with straw, dressed in old uniforms, and that a mere handful of men were driving the animals forward. The cavalrymen had purposely dismounted and secreted themselves in the wood in anticipation of such a pursuit as was made.

While the Germans do not appear to be so enterprising in this form of ingenuity they have not been idle. A French airman flying over the Teuton lines observed the outermost trenches to be alive with men whose helmets were distinctly visible. The airman reported his observations and the trench was subjected to terrific shell fire. Subsequently the French made a spirited charge, but to their dismay found that the outermost German trench was occupied by dummies fashioned from all sorts of materials and crowned with helmets! This ruse had enabled the German lines to be withdrawn to another position in safety and comparatively at leisure.

Before war was declared the German military experts were emphasising the importance of trees for masking troops and guns against aerial observation. One of the foremost authorities upon military aviation only a few months ago urged the German Military Staff to encourage the planting of orchards, not for the purpose of benefiting agriculture or in the interests of the farmers, but merely for military exigencies.

He pointed to the extensive orchards which exist in Alsace-Lorraine and Baden, the military covering value of which he had determined from personal experience, having conducted aerial operations while military were moving to and fro under the cover of the trees. He declared that the cover was efficient and that under the circumstances the laying out of extensive orchards in strategical places should be carried out without any delay. This, he urged, was a national and not a private obligation. He advocated the bestowal of subsidies on the farmers to encourage the planting of fruit trees. He suggested that the trees should be provided by the State, and given to all who were prepared to plant them; that substantial prizes should be awarded to encourage the rapid growth thereof, and that annual prizes should be awarded to the man who would undertake their cultivation and pruning, not from the fruit-yielding point of view, but for facilitating the movement of troops beneath their dense branches.

He even urged the military acquisition of suitable land and its determined, skilful, and discreet exploitation by those who loved the Fatherland. He emphasised the necessity for keeping such orchards under military control, only vouchsafing sufficient powers to the local authorities to ensure the desired consummation. He maintained that, if the work were prosecuted upon the right lines and sufficient financial assistance were given, the purpose in view could be achieved without saddling the war department with any unremunerative or excessive burden. He admitted that the process of raising fruit trees to the stage when they would afford adequate cover would be tedious and somewhat prolonged, but argued that the military advantages, such as enabling troops to move below the welcome shelter with absolute freedom and without physical fatigue, would be an ample compensation.

The utility of such cover to artillery was another factor he did not fail to emphasise. He dwelt seriously upon the difficulty of rendering permanent gun emplacements and heavy artillery invisible to the airman by resort to the usual type of gun shields. The latter may be located with ease by alert airmen, whereas if the guns were under cover of fruit trees they would be able to accomplish their deadly mission without betraying their presence to the aerial scout. Moreover, by pruning the trees in such a manner as to ensure free movement beneath, the artillery would be able to advance without betraying the fact to the enemy.

This authority vigorously insisted that the work should be carried out without a moment's delay as it was vital to the Fatherland. In the light of recent events, and the excellent cover which is offered by the orchards of the territory he cited as an illustration of his contention, such a disclosure is pregnant with meaning. It throws a new light upon the thorough methods with which the Germans carried out their military preparations, and incidentally shows that they were fully alive to every possible development. Fruit-raising as a complement to military operations may be a new line of discussion, but it serves to reveal the German in his true light, ready for every contingency, and shows how thoroughly he appreciates the danger from the man in the clouds.

CHAPTER XIV

ANTI-AIRCRAFT GUNS. MOBILE WEAPONS.

When the airship and the aeroplane became accepted units of warfare it was only natural that efforts should be concentrated upon the evolution of ways and means to compass their destruction or, at least, to restrict their field of activity. But aircraft appeared to have an immense advantage in combat. They possess virtually unlimited space in which to manoeuvre, and are able to select the elevation from which to hurl their missiles of destruction.

There is another and even more important factor in their favour. A projectile fired, or even dropped, from a height, say of 5,000 feet, is favourably affected by the force of gravity, with the result that it travels towards the earth with accumulating energy and strikes the ground with decisive force.

On the other hand, a missile discharged into space from a weapon on the earth has to combat this action of gravity, which exercises a powerful nullifying influence upon its flight and velocity, far in excess of the mere resistance offered by the air. In other words, whereas the projectile launched from aloft has the downward pull of the earth or gravitational force in its favour, the shell fired from the ground in the reverse direction has to contend against this downward pull and its decelerating effect.

At the time when aircraft entered the realms of warfare very little was known concerning the altitudes to which projectiles could be hurled deliberately. Certain conclusive information upon this point was available in connection with heavy howitzer fire, based on calculations of the respective angles at which the projectile rose into the air and fell to the ground, and of the time the missile took to complete its flight from the gun to the objective. But howitzer fire against aircraft was a sheer impossibility: it was like using a six-inch gun to kill a fly on a window pane at a thousand yards' range. Some years ago certain experiments in aerial firing with a rifle were undertaken in Switzerland. The weapon was set vertically muzzle upwards and discharged. From the time which elapsed between the issue of the bullet from the muzzle until it struck the earth it was possible to make certain deductions, from which it was estimated that the bullet reached an altitude of 600 feet or so. But this was merely conjecture.

Consequently when artillerists entered upon the study of fighting air-craft with small arms and light guns, they were compelled to struggle in the dark to a very pronounced extent, and this darkness was never satisfactorily dispelled until the present war, for the simple reason that there were no means of getting conclusive information. The German armament manufacturers endeavoured to solve the problem by using smoking shells or missiles fitted with what are known as tracers. By following the ascensional path of the projectiles as revealed by the smoke it was possible to draw certain conclusions. But these were by no means convincing or illuminating, as so many factors affected the issue.

Despite the peculiar and complex difficulties associated with the problem it was attacked some what boldly. In this trying field of artillery research the prominent German armament manufacturers, Krupp of Essen and Ehrhardt of Dusseldorf, played a leading part, the result being that before the airship or the aeroplane was received within the military fold, the anti-aircraft gun had been brought into the field of applied science. The sudden levelling-up serves to illustrate the enterprise of the Germans in this respect as well as their perspicacity in connection with the military value of aircraft.

Any gun we can hope to employ against aircraft with some degree of success must fulfil special conditions, for it has to deal with a difficult and elusive foe. Both the lighter-than-air and the heavier than-air craft possess distinctive features and varying degrees of mobility. Taking the first-named, the facility with which it can vary its altitude is a disconcerting factor, and is perplexing to the most skilful gunner, inasmuch as he is called upon to judge and change the range suddenly.

On the other hand, the artilleryman is favoured in certain directions. The range of utility of the airship is severely limited. If its avowed mission is reconnaissance and conclusive information concerning the disposition of forces, artillery and so forth is required, experience has proved that such work cannot be carried out satisfactorily or with any degree of accuracy at a height exceeding 5,000 feet, and a distance beyond six miles. But even under these circumstances the climatic conditions must be extremely favourable. If the elements are unpropitious the airship must venture nearer to its objective. These data were not difficult to collect, inasmuch as they were more or less available from the results of military observations with captive balloons, the conditions being somewhat similar. With the ordinary captive balloon it has been found that, in clear weather, a radius of about 3 3/4 miles at the maximum elevation constitutes its range of reliable utility.

With the aeroplane, however, the conditions are very dissimilar. In the first place the machine owing to its diminutive size as compared with the airship, offers a small and inconspicuous target. Then there is its high independent speed, which is far beyond that of the airship. Furthermore its mobility is greater. It can wheel, turn sharply to the right or to the left, and pursue an irregular undulating flight in the horizontal plane, which renders it well nigh impossible for a gunner to pick it up. The machine moves at a higher relative speed than that at which the gun can be trained. It is the rapid and devious variation which so baffles the gunner, who unless he be highly skilled and patient, is apt to commence to fire wildly after striving for a few moments, and in vain, to pick up the range; he trusts to luck or depends upon blind-shooting, which invariably results in a waste of ammunition.

A gun, to be of tangible destructive efficiency when directed against aircraft, especially those depending upon the gas-bag for equilibrium, must be of special design. It must be capable of firing at an angle only a few degrees less than the absolute vertical, and in order to follow the rapid and involved movements of its objective, must be so mobile that it can be trained through a complete circle at any angle of inclination less than its maximum. At the same time, if the weapon is being used in field operations it must be mounted upon a carriage of adequate mobility to enable it to follow the airship, and thereby keep pace with the latter, so that the aerial craft may be sorely harassed if not actually hit. The automobile is the obvious vehicle for this duty, and it has accordingly been extensively used in this service.

The automobile and the gun mounted thereon follow widely different lines. Some vehicles are designed especially for this duty, while others are improvisations, and be it noted, in passing, that many of the latter have proved more serviceable than the former. Still, the first-named is to be preferred, inasmuch as necessarily it is designed to meet the all-round requirements imposed, and consequently is better able to stand up to the intended work, whereas the extemporised vehicle is only serviceable under favourable conditions.

The Krupp Company has evolved many designs of anti-aircraft motor-driven guns--"Archibalds" the British airmen term them with emphatic levity. They are sturdily-built vehicles fitted with heavy motors, developing from 40 to 50 horse-power, with the chassis not widely dissimilar from that adopted for motor-omnibus traffic. Consequently, they are not necessarily condemned to the high-roads, but within certain limits are able to travel across country, i.e., upon fields or other level expanses, where the soil is not unduly soft.

But the very character of the problem rendered the evolution of the vehicle a somewhat perplexing matter. There were many factors which had to be taken into consideration, and it was possible to meet the imposed requirements only within certain limits. In the first place, the weight of the gun itself had to be kept down. It was obviously useless to overload the chassis. Again, the weight of the projectile and its velocity had to be borne in mind. A high velocity was imperative. Accordingly, an initial velocity varying from 2,200 to 2,700 feet per second, according to the calibre of the gun, was determined.

Moreover, as mobility was an indispensable condition, the gun had to be so mounted that it could be fired from the motor-car even if the latter were travelling at high speed. This requirement entailed another difficulty. The gun had to be mounted in such a manner as to enable the gunner to train it easily and readily through the complete circle and through its complete range of vertical inclination. As the result of prolonged experiments it was ascertained that the most suitable arrangement was a pedestal mounting, either within a turret or upon an open deck. To meet the weight of the gun, as well as the strains and stresses incidental to firing, the chassis was strengthened, especially over the rear axle near which the mounting is placed.

The heaviest gun of this type is the 10.5 centimetre (4 1/4-inch) quick-firer, throwing a shell weighing nearly forty pounds, with an initial velocity of 2,333 feet per second. This "Archibald" is totally unprotected. The gun is mounted centrally upon the carriage over the rear axle, and occupies the centre of the deck between the driver's seat and that of the gun crew behind. The whole of the deck is clear, thereby offering no obstruction to the gunner in training the weapon, while the space may be widened by dropping down the wings of the vehicle. At the rear is a seat to accommodate the gun crew, beneath which the ammunition is stowed. When travelling and out of action, the gun lies horizontally, the muzzle pointing from the rear of the car.

To reduce the strains arising from firing, the arm is fitted with what is known as the "differential recoil." Above the breach is an air recuperator and a piston, while there is no hydraulic brake such as is generally used. The compressor is kept under compression while the car is travelling with the gun out of action, so that the arm is available for instant firing. This is a departure from the general practice in connection with such weapons. When the gun is loaded the bolt which holds the compressor back is withdrawn, either by the hand for manual firing, or by the action of the automatic closing of the breech when the arm is being used as a quick-firer. In firing the gun is thrown forward under the pressure of the released air which occurs at the moment of discharge. The energy of the recoil brings the gun back and at the same time recharges the compressed air reservoir.

The gun is so mounted upon its pedestal as to enable a maximum vertical inclination of 75 degrees to be obtained. The mounting system also enables the weapon to be trained in any desired direction up to the foregoing maximum elevation throughout a complete circle, and it can be handled with ease and celerity. A smaller "Archibald" is the 7.5 centimetre (3-inch gun) throwing a 14.3 pound shell at an initial velocity of about 2,170 feet per second.

The turret anti-aircraft gun carried upon a motor-car differs from the foregoing very considerably. This is a protected arm. The gun of 7.1 centimetres--approximately 2.75 inches--is mounted in the same manner upon the car-deck and over the driving axle, but is enclosed within a sheet steel turret, which is proof against rifle and machine-gun fire. This turret resembles the conning-tower of a battleship, and is sufficiently spacious to house the whole of the gun crew, the internal diameter being about seven feet. Access to the turret is obtained through a rear door. This gun has a maximum elevation of about 75 degrees, while its operation and mechanism are similar to those of the unprotected weapon.

The vehicle itself is practically identical with the armoured motor-car, which has played such an important part during the present campaign, the driver being protected by a bullet-proof steel screen similar in design to the ordinary glass wind-screen fitted to touring automobiles. This is carried sufficiently high to offer complete protection to his head when seated at the wheel, while through a small orifice in this shield he is able to obtain a clear view of the road. The engine and its vital parts are also adequately protected. The ammunition is carried in a cupboard-like recess forming part of the driver's seat, encased in bullet-proof steel sheeting with flap-doors. This device enables the shells to be withdrawn readily from the side of the car and passed to the crew within the turret. The caisson is of sufficient dimensions to receive 69 shells.

The Ehrhardt airship fighting ordnance is similarly adapted to motor-car operations, one type being especially powerful. The whole of the vehicle is encased in armour-plating impervious to rifle and machine-gun fire. The driver is provided with a small orifice through which he is able to obtain a clear uninterrupted view of the road ahead, while the armouring over the tonneau is carried to a sufficient height to allow head-room to the gun crew when standing at the gun. All four wheels are of the disk type and fashioned from heavy sheet steel. The motor develops 40-50 horse-power and, in one type, in order to mitigate the risk of breakdown or disablement, all four wheels are driven. The gun, a small quick-firer, is mounted on a pedestalin a projecting conning-tower. The mounting is placed behind the driver's seat, and is trained and operated from the tonneau. The maximum elevation is 75 degrees, and like the gun carriage bearing the tube guide it can be moved through a complete circle, being free to rotate in the fixed pivot jack to enable this end to be attained.

The foregoing may be said to represent the most powerful types of mobile anti-aircraft weapons used by the Austro-German forces to-day. Arms of similar design, roughly speaking, have also been introduced into the French and Russian services. In addition many semi-armoured weapons of this character are in operation, some specially built for the work, while others have been improvised. In the semi-armoured motor-car the carriage follows the usual lines; it has an open top, the armouring comprising the body of the tonneau and the diskwheels, which are made of light bullet-proof steel. Here again the prevailing practice is to mount the gun as nearly above the rear axle as possible, and to work it from the tonneau. The maximum elevation is also

75 degrees, with training throughout the entire circle.

Another type comprises a very light machine gun of rifle calibre, and this is intended for attachment to an ordinary motor car. There is a pedestal mounting which can be set within the tonneau, while the weapon is pivoted in an outrigger, the latter being free to rotate in its pivot jack. This arrangement enables the arm to cover a wide range,while it also admits of training through an extensive angle of elevation.

The Allied forces improvised travelling anti-aircraft offences by mounting the latest types of Vickers, Hotchkiss, and other machine guns in armoured motor cars. Some of these have the domed turret form, with the gun projecting through the roof, while others are protected against hostile attack from the side only, the carriage being panelled with bullet-proof steel sheeting. While such weapons are useful, inasmuch as they can maintain a hot fire ranging up to 750 shots per minute, they are not to be compared with the "Archibalds," which are able to throw heavy shrapnel and incendiary shells, and have a vertical range of about 6,000 to 8,000 feet.

The improvised motor-gun has not proved a complete success, except in those instances when the hostile aircraft has ventured to approach somewhat closely to the ground. The more formidable weapons cannot be mounted upon ordinary vehicles, inasmuch as the increase in weight, which is appreciable, impairs the efficiency of the vehicle, and at the same time enhances the possibility of breakdown at a critical moment. For such arms a special and substantial chassis is imperative, while the motive power and gearing must be adapted to the circumstances.

Motor-mounted anti-aircraft weapons, however, have not proved an unqualified success. The fact that the vehicles are condemned to the high roads, or at least to comparatively smooth and level ground, constitutes a severe handicap. Again, when travelling at high speed, and this is essential when pursuing a fast aeroplane, the accurate laying of the weapon is extremely difficult, owing to the oscillation of the vehicle itself, especially if the road surface is in a bad condition. The sighting arrangements are of a wonderfully complete character, as described elsewhere, but the irregular rolling movement arising from high speed is a nullifying quantity. It is tolerably easy for the aircraft, especially an aeroplane, to evade successful pursuit, either by rising to an elevation beyond the range of the gun, or by carrying out baffling evolutions such as irregular undulating flight, wheeling, and climbing. According to the reports of the British and French airmen the "Archibald" has failed to establish the glowing reputation which was anticipated, for the simple reason that, unless it has a clear straight road and can maintain its high speed, it can easily be out-distanced by the fleet human bird.

The motor-car suffers from another serious disability. It cannot manoeuvre with sufficient celerity. For instance, if it is necessary to turn round in a narrow lane, valuable time is lost in the process, and this the airman turns to account. In hilly country it is at a still greater disadvantage, the inclines, gradients, and sinuosities of the roads restricting its effectiveness very pronouncedly. It must also be remembered that, relatively speaking, the "Archibald" offers a better target to the airman than the aeroplane offers to the man behind the anti-aircraft gun on the motor below. A few well-placed bombs are sufficient to induce the pursuers to cease their activities. Even if the missiles fail to strike the motor-car itself they can wreak disaster in directly by rendering the road impassable or dangerous to negotiate at high speed. On the whole therefore, the "Archibald" is a greatly exaggerated weapon of offence against aircraft, and, so far as is known, has failed to fulfil expectations. In fact, the Germans have practically abandoned the idea of using it in the manner of a pursuing arm; they work the weapon as a fixture, depending upon the car merely as a means of moving it from point to point. Thus, in reality, it has been converted into a light field-piece, and may almost be included in the category of fixed weapons for combating aerial operations.

CHAPTER XV

ANTI-AIRCRAFT GUNS. IMMOBILE WEAPONS

The immobile anti-aircraft gun, as distinct from that attached to a travelling carriage such as a motor-car, may be subdivided into two classes. The one is the fixed arm which cannot be moved readily, mounted upon a permanent emplacement; the other is the field-piece which, while fired from a stationary position, may be moved from point to point upon a suitable carriage. The distinction has its parallel in ordinary artillery, the first-named weapon coinciding with the heavy siege gun, which is built into and forms part and parcel of the defensive or offensive scheme, while the second is analogous to the field artillery, which may be wheeled from position to position.

In this phase of artillery the Germans led the way, for the simple reason that they recognised the military value of aerial navigation years in advance of their contemporaries. Again, in this field the Krupp Organisation has played a prominent part. It embarked upon actual construction of weapons while its rivals in other countries were content to prepare their drawings, which were filed against "The Day." But it must not be thought that because the German manufacturers of armaments were ahead of their contemporaries they dominated the situation. Far from it. Their competitors in the market of destruction were every whit as keen, as ingenious, and as enterprising. Kruppism saw a commercial opportunity to profit from advertisement and seized it: its rivals were content to work in secret upon paper, to keep pace with the trend of thought, and to perfect their organisations so as to be ready for the crisis when it developed.

The first Krupp anti-aircraft field-piece was a 6.5 centimetre (2 9/16 inch) arm. It possessed many interesting features, the most salient of which was the design of the axle of the carriage. The rigid axle for the two wheels was replaced by an axle made in two sections, and joined together in the form of a universal coupling, so that each wheel virtually possessed its own axle, or rather half-axle. This was connected with the cradle of the gun in such a manner that the wheels were laterally pivoted thereon.

The result is that each axle can be turned forward together with its wheel, and thus the wheels have their rims brought into line to form an arc of a circle, of which the rear end of the spade of the gun carriage constitutes the centre. This acts as a pivot, about which the gun can be turned, the pair of wheels forming the runners for the achievement of this movement. The setting of the weapon in the firing position or its reversion to the travelling position can be easily and speedily effected merely by the rotation of a handwheel and gearing.

With this gun a maximum elevation of 60 degrees is possible, owing to the trunnions being carried well behind the breech in combination with the system of long steady recoil. The balancing spring which encloses the elevating screw is contained in a protected box. The recoil brake, together with the spring recuperator, follows the usual Krupp practice in connection with ordinary field pieces, as does also the automatic breech-closing and firing mechanism. In fact there is no pronounced deviation from theprevailing Krupp system, and only such modifications as are necessary to adapt the arm to its special duty. When the gun is elevated to high angles the shell, after insertioin the breech, is prevented from slipping out by means of a special device, so that the proper and automatic closing of the breech is not impaired in any way.

In such an arm as this, which is designed essentially for high-angle firing, the sighting and training facilities require to be carried out upon special lines, inasmuch as the objective is necessarily at a considerable altitude above the horizon of the gun. In other words, in firing at a high inclination, distance between the gun and the target cannot be utilised directly for the back sight. On the other hand, it is essential that in proportion as the angle from the horizontal increases, the back sight should be lowered progressively in a manner corresponding to the distance.

To assist the range-finder in his task of sighting it is necessary that he should be provided with firing tables set out in a convenient form, which, in conjunction with the telemeter, serve to facilitate training for each successive round. In this way it is possible to pick up the range quickly and to keep the objective in the line of fire until it either has been put hors de combat, or has succeeded in retiring beyond the range of the gun.

The sighting arrangements of these Krupp anti-aircraft guns are carried out upon these lines. Beneath the barrel of the back-sight is an observing glass with an eye-piece for the artillerist, while above and behind the observing glass is another eye-piece, to be used in conjunction with the manipulation of the back-sight. The eye-piece of the observation glass is so made that it can be turned through a vertical plane in proportion as the angle of fire increases in relation to the horizontal. The determination of the distance from the objective and from the corresponding back-sight as well as the observation of the altitude is carried out wlth the aid of the telemeter. This again carries an observation glass fitted with an eye-piece which can be turned in the vertical plane in the same manner as that of the fore-sight. By means of this ingenious sighting device it is possible to ascertain the range and angle of fire very easily and speedily.

The weight of the special Krupp anti-aircraft field-piece, exclusive of the protecting shield, is approximately identical with that of the ordinary light artillery field-piece. It throws a shell weighing 8.8 pounds with an initial velocity of about 2,066 feet per second.

Although the German armament manufacturers were among the first to enter the field with an anti-aircraft gun of this character they were speedily followed by the French, who devised a superior weapon. In fact, the latter represented such a decisive advance that the German artillerists did not hesitate to appropriate their improvements in sundry essential details, and to incorporate them with their own weapons. This applies especially to the differential recoil system which is utilised in the small anti-aircraft guns now mounted upon the roofs of high buildings of cities throughout Germany for the express purpose of repelling aerial attack.

The French system is admitted by the leading artillery technicians of the world to be the finest which has ever been designed, its remarkable success being due to the fact that it takes advantage of the laws of Nature. In this system the gun is drawn back upon its cradle preparatory to firing. In some instances the barrel is compressed against a spring, but in the more modern guns it is forced to rest against a cushion of compressed air contained within a cylinder. When first bringing the gun into action, the barrel is brought into the preliminary position by manually compressing the air or spring by means of a lever. Thereafter the gun works automatically. When the gun is fired the barrel is released and it flies forward. At a critical point in its forward travel the charge is fired and the projectile speeds on its way. The kick or recoil serves to arrest the forward movement of the barrel and finally drives it back again against the strong spring or cushion of compressed air within the cylinder to its normal position, when it is ready for the introduction of the next shell.

The outstanding feature of this system is that the projectile is given a higher initial velocity than is possible with the barrel held rigid at the moment of discharge, because the shell is already travelling at the moment of firing.

The fixed anti-aircraft guns such as are stationed upon eminences and buildings are of the quick firing type, the object being to hurl a steady, con tinuous stream of missiles upon the swiftly moving aeroplane. Some of the weapons throw a one-pound shell and are closely similar to the pom-pom which proved so effective during the South African war. Machine guns also have been extensively adopted for this duty by all the combatants, their range of approximately 2,000 yards and rapidity of fire being distinctly valuable when hostile aircraft descend to an altitude which brings them within the range of the weapon.

The greatest difficulty in connection with this phase of artillery, however, is not so much the evolution of a serviceable and efficient type of gun, as the determination of the type of projectile which is likely to be most effective. While shrapnel is employed somewhat extensively it has not proved completely satisfactory. It is difficult to set the timing fuse even after the range has been found approximately, which in itself is no easy matter when the aircraft is moving rapidly and irregularly, but reliance is placed thereon in the hope that the machine may happen to be within the cone of dispersion when the shell bursts, and that one or more of the pieces of projectile and bullets may chance to penetrate either the body of the airman or a vital part of the mechanism.

It is this uncertainty which has led to a preference for a direct missile such as the bullet discharged from a machine gun. A stream of missiles, even of rifle calibre, maintained at the rate of some 400 shots per minute is certain to be more effective, provided range and aim are correct, than shrapnel. But the ordinary rifle-bullet, unless the objective is within very close range, is not likely to cause much harm, at least not to the mechanism of the aerial vessel.

It is for this reason that greater attention is being devoted, especially by the French artillerists, to the Chevalier anti-aircraft gun, a weapon perfected by a Swiss technician resident in Great Britain. It projects a formidable missile which in fact is an armour-piercing bullet 1/2- to 3/4-inch in diameter. It is designed for use with an automatic machinegun, which the inventor has devised more or less upon the well-known French system. The bullet has a high velocity--about 2,500 feet per second--and a maximum range of 6,000 to 8,000 feet at the maximum elevation. Should such a missile strike the motor or other mechanism of the vessel it would wreak widespread havoc, and probably cause the machine to come to earth. This arm has been designed for the express purpose of disabling the aeroplane, and not for the subjugation of the airman, which is a minor consideration, inasmuch as he is condemned to a descent when his craft receives a mortal wound.

Attempts have been and still are being made to adapt an explosive projectile to this gun, but so far the measure of success achieved has not proved very promising. There are immense difficulties connected with the design of an explosive shell of this class, charged with a high explosive, especially in connection with the timing. So far as dependence upon percussive detonation is concerned there is practically no difficulty. Should such a missile strike, say, the motor of an aeroplane, or even the hull of the craft itself, the latter would be practically destroyed. But all things considered, it is concluded that more successful results are likely to be achieved by the armour-piercing bullet striking the mechanism than by an explosive projectile.

The Krupp company fully reahsed the difficulties pertaining to the projectile problem in attacks upon aerial craft. So far as dirigibles are concerned shrapnel is practically useless, inasmuch as even should the bag be riddled by the flying fragments, little effective damage would be wrought--the craft would be able to regain its haven. Accordingly efforts were concentrated upon the perfection of two new types of projectiles, both of which were directed more particularly against the dirigible. The one is the incendiary shell--obus fumigene--while the other is a shell, the contents of which, upon coming into contact with the gas contained within the gas-bag, set up certain chemical reactions which precipitate an explosion and fire.

The incendiary shells are charged with a certain compound which is ignited by means of a fuse during its flight. This fuse arrangement coincides very closely with that attached to ordinary shrapnel, inasmuch as the timing may be set to induce ignition at different periods, such as either at the moment it leaves the gun, before, or when it strikes the envelope of the dirigible. The shell is fitted with a "tracer," that is to say, upon becoming ignited it leaves a trail of smoke, corresponding with the trail of a rocket, so that its passage through the air may be followed with facility. This shell, however, was designed to fulfil a dual. Not only will it fire the gaseous contents out of the dirigible, but it has an explosive effect upon striking an incombustible portion of the aircraft, such as the machinery, propellers or car, when it will cause sufficient damage to throw the craft out of action.

The elaborate trials which were carried out with the obus fumigene certainly were spectacular so as they went. Two small spherical balloons, 10 feet in diameter, and attached to 1,000 feet of cable, were sent aloft. The anti-aircraft guns themselves were placed about 5,100 feet distant. Owing to the inclement weather the balloons were unable to attain a height of more than 200 feet in a direct vertical line above the ground. The guns were trained and fired, but the one balloon was not hit until the second round, while the third escaped injury until the fifth round. When struck they collapsed instantly. Though the test was not particularly conclusive, and afforded no reliable data, one point was ascertained--the trail of smoke emitted by the shell enabled its trajectory to be followed with ease. Upon the conclusion of these trials, which were the most successful recorded, quick-firing tests in the horizontal plane were carried out. The best performance in this instance was the discharge of five rounds in eight seconds. In this instance the paths of the projectiles were simple and easy to follow, the flight of the shell being observed until it fell some 18,670 feet away. But the Krupp firmhave found that trials upon the testing ground with a captive balloon differ very materially from sterntests in the field of actual warfare. Practically nothing has been heard of the two projectiles during this war, as they have proved an absolute failure.

Some months ago the world was startled by the announcement that the leading German armament firm had acquired the whole of the interest in an aerial torpedo which had been evolved by the Swedish artillerist, Gustave Unge, and it was predicted that in the next war widespread havoc would be wrought therewith. Remarkable claims were advanced for this projectile, the foremost being that it would travel for a considerable distance through the air and alight upon the objective with infallible accuracy. The torpedo in question was subjected to exacting tests in Great Britain, which failed to substantiate all the claims which were advanced, and it is significant to observe that little has been heard of it during the present conflict. It is urged in certain technical quarters, however, that the aerial torpedo will prove to be the most successful projectile that can be used against aircraft. I shall deal with this question in a later chapter.

During the early days of the war anti-aircraft artillery appeared to be a much overrated arm. The successes placed to its credit were insignificant. This was due to the artillerymen being unfamiliar with the new arm, and the conditions which prevail when firing into space. Since actual practice became possible great advances in marksmanship have been recorded, and the accuracy of such fire to-day is striking. Fortunately the airman possesses the advantage. He can manoeuvre beyond the range of the hostile weapons. At the moment 10,000 feet represents the extreme altitude to which projectiles can be hurled from the arms of this character which are now in use, and they lack destructiveness at that range, for their velocity is virtually expended.

Picking up the range is still as difficult as ever. The practice followed by the Germans serves to indicate the Teuton thoroughness of method in attacking such problems even if success does not ensue. The favourite German principle of disposing anti-aircraft artillery is to divide the territory to be protected into equilateral triangles, the sides of which have a length of about six miles or less, according to the maximum effective range of the pieces at an elevation of 23 1/2 degrees.

The guns are disposed at the corners of the triangles as indicated in Figs. 13-14. Taking the one triangle as an example, the method of picking up the range may be explained as follows. The several guns at the comers of the triangle, each of which can be trained through the 360 degrees in the horizontal plane, are in telephonic touch with an observer O stationed some distance away. The airman A enters the area of the triangle. The observer takes the range and communicates with the gunner B, who fires his weapon. The shell bursts at 1 emitting a red flame and smoke. The observer notes the altitude and relative position of the explosion in regard to the aircraft, while gunner B himself observes whether the shell has burst to the right or to the left of the objective and corrects accordingly. The observer commands C to fire, and another shell is launched which emits a yellow flame and smoke. It bursts at 2 according to the observer, while gunner C also notes whether it is to the right or to the left of the target and corrects accordingly. Now gunner D receives the command to fire and the shell which explodes at 3 throws off a white flame and smoke. Gunner D likewise observes whether there is any deviation to right or left of the target and corrects in a similar manner. From the sum of the three rounds the observer corrects the altitude, completes his calculations, and communicates his instructions for correction to the three gunners, who now merely train their weapons for altitude. The objective is to induce the shells hurled from the three corners of the triangle to burst at a common point 4, which is considered to be the most critical spot for the aviator. The fire is then practically concentrated from the three weapons upon the apex of a triangular cone which is held to bring the machine within the danger zone.

This method of finding the range is carried out quickly--two or three seconds being occupied in the task. In the early days of the war the German anti-aircraft artillerymen proved sadly deficient in this work, but practice improved their fire to a marvellous degree, with the result that at the moment it is dangerous for an aviator to essay his task within an altitude of 6,000 feet, which is the range of the average anti-aircraft gun.

The country occupied by a belligerent is divided up in this manner into a series of triangles. For instance, a machine entering hostile territory from the east, enters the triangle A-B-C, and consequently comes within the range of the guns posted at the comers of the triangle. Directly he crosses the line B-C and enters the adjacent triangle he passes beyond the range of gun A but comes within the range of the gun posted at D, and while within the triangular area is under fire from the guns B-C-D. He turns and crosses the line A-C, but in so doing enters another triangle A-C-E, and comes range of the gun posted at E.

The accompanying diagram represents an area of country divided up into such triangle and the position of the guns, while the circle round the latter indicate the training arc of the weapons, each of which is a complete circle, in the horizontal plane. The dotted line represents the aviator's line of flight, and it will be seen that no matter how he twists and turns he is always within the danger zone while flying over hostile territory. The moment he outdistances one gun he comes within range of another.

The safety of the aviator under these circumstances depends upon his maintaining an altitude exceeding the range of the guns below, the most powerful of which have a range of 8,000 to 10,000 feet, or on speed combined with rapid twisting and turning, or erratic undulating flight, rendering it extremely difficult for the gun-layer to follow his path with sufficient celerity to ensure accurate firing.

At altitudes ranging between 4,000 and 6,000 feet the aeroplane comes within the range of rifle and machine-gun firing. The former, however, unless discharged in volleys with the shots covering a wide area, is not particularly dangerous, inasmuch as the odds are overwhelmingly against the rifleman. He is not accustomed to following and firing upon a rapidly moving objective, the result being that ninety-nine times out of a hundred he fails to register a hit. On the other hand the advantage accruing from machine-gun fire is, that owing to the continuous stream of bullets projected, there is a greater possibility of the gun being trained upon the objective and putting it hors de combat.

But, taking all things into consideration, and notwithstanding the achievements of the artillerist, the advantages are overwhelmingly on the side of the aviator. When one reflects upon the total sum of aircraft which have been brought to earth during the present campaign, it will be realised that the number of prizes is insignificant in comparison with the quantity of ammunition expended.

CHAPTER XVI

MINING THE AIR

While the anti-aircraft gun represents the only force which has been brought to the practical stage for repelling aerial attack, and incidentally is the sole offensive weapon which has established its effectiveness, many other schemes have been devised and suggested to consummate these ends. While some of these schemes are wildly fantastic, others are feasible within certain limitations, as for instance when directed against dirigibles.

It has been argued that the atmosphere is akin to the salt seas; that an aerial vessel in its particular element is confronted with dangers identical with those prevailing among the waters of the earth. But such an analogy is fallacious: there is no more similarity between the air and the ocean than there is between an airship and a man-of-war. The waters of the earth conceal from sight innumerable obstructions, such as rocks, shoals, sandbanks, and other dangers which cannot by any means be readily detected.

But no such impediments are encountered in the ether. The craft of the air is virtually a free age in the three dimensions. It can go whither it will without let or hindrance so long as the mechanical agencies of man are able to cope with the influences of Nature. It can ascend to a height which is out of all proportion to the depth to which the submarine can descend in safety. It is a matter of current knowledge that a submarine cannot sink to a depth of more than 250 feet: an aerial vessel is able to ascend to 5,000, 8,000, or even 10,000 feet above the earth, and the higher the altitude it attains the greater is its degree of safety. The limit of ascension is governed merely by the physical capacities of those who are responsible for the aerial vessel's movement.

It is for this reason that the defensive measures which are practised in the waters of the earth are inapplicable to the atmosphere. Movement by, or in, water is governed by the depth of channels, and these may be rendered impassable or dangerous to negotiate by the planting of mines. A passing ship or submarine may circumvent these explosive obstructions, but such a successful manoeuvre is generally a matter of good luck. So far as submarines are concerned the fact must not be over looked that movements in the sea are carried out under blind conditions: the navigator is unable to see where he is going; the optic faculty is rendered nugatory. Contrast the disability of the submarine with the privileges of its consort in the air. The latter is able to profit from vision. The aerial navigator is able to see every inch of his way, at least during daylight. When darkness falls he is condemned to the same helplessness as his confrere in the waters below.

A well-known British authority upon aviation suggested that advantage should be taken of this disability, and that the air should be mined during periods of darkness and fog to secure protection against aerial invasion. At first sight the proposal appears to be absolutely grotesque, but a little reflection will suffice to demonstrate its possibilities when the area to be defended is comparatively limited. The suggestion merely proposes to profit from one defect of the dirigible. The latter, when bent upon a daring expedition, naturally prefers to make a bee-line towards its objective: fuel considerations as a matter of fact compel it to do so. Consequently it is possible, within certain limits, to anticipate the route which an invading craft will follow: the course is practically as obvious as if the vessel were condemned to a narrow lane marked out by sign-posts. Moreover, if approaching under cover of night or during thick weather, it will metaphorically "hug the ground." To attempt to complete its task at a great height is to court failure, as the range of vision is necessarily so limited.

Under these circumstances the mining of the air could be carried out upon the obvious approaches to a threatened area. The mines, comprising large charges of high-explosive and combustible material, would be attached to small captive balloons similar to the "sounding balloons" which are so much used by meteorologists in operations for sounding the upper strata of the atmosphere. These pilot balloons would be captive, their thin wires being wound upon winches planted at close intervals along the coast-line. The balloon-mines themselves would be sent to varying heights, ranging from 1,000 to 5,000 feet, and with several attached to each cable, the disposition of the mines in the air in such an irregular manner being in fact closely similar to the practice adopted in the mining of a channel for protection against submarines and hostile ships.

The suggestion is that these mines should be sent aloft at dusk or upon the approach of thick and foggy weather, and should be wound in at dawn or when the atmosphere cleared, inasmuch as in fine weather the floating aerial menace would be readily detected by the pilot of a dirigible, and would be carefully avoided. If the network were sufficiently intricate it would not be easy for an airship travelling at night or in foggy weather to steer clear of danger, for the wires holding the balloons captive would be difficult to distinguish.

The mines would depend upon detonators to complete their work, and here again they would bear a close resemblance to sea-mines. By looping the mines their deadliness could be increased. The unsuspicious airship, advancing under cover of darkness or thick weather, might foul one of the wires, and, driving forward, would tend to pull one or more mines against itself. Under the force of the impact, no matter how gentle, or slight, one or more of the detonating levers would be moved, causing the mine to explode, thus bursting the lifting bag of the vessel, and firing its gaseous contents. An alternative method, especially when a cable carried only a single mine, would be to wind in the captive balloon directly the wire was fouled by an invading aerial craft, the process being continued until the mine was brought against the vessel and thereby detonated.

Another proposed mining method differs materially in its application. In this instance it is suggested that the mines should be sent aloft, but should not be of the contact type, and should not be fired by impact detonators, but that dependence should be placed rather upon the disturbing forces of a severe concussion in the air. The mines would be floating aoft, and the advance of the airship would be detected. The elevation of the mines in the vicinity of the invading craft would be known, while the altitude of the airship in relation thereto could be calculated. Then, it is proposed that a mine within d certain radius of the approaching craft, and, of course, below it, should be fired electrically from the ground. It is maintained that if the charge were sufficiently heavy and an adequate sheet of flame were produced as a result of the ignition, an airship within a hundred yards thereof would be imperilled seriously, while the other mines would also be fired, communicating ignition from one to the other. The equilibrium of the airship is so delicate that it can be readily upset, and taking into account the facts that gas is always exuding from the bag, and that hydrogen has a tendency to spread somewhat in the manner of oil upon water, it is argued that the gas would be ignited, and would bring about the explosion of the airship.

Another method has even been advocated. It is averred in authoritative circles that when the aerial invasion in force of Great Britain is attempted, the Zeppelins will advance under the cover of clouds. Also that the craft will make for one objective--London. Doubtless advantage will be taken of clouds, inasmuch as they will extend a measure of protection to the craft, and will probably enable the invading fleet to elude the vigilance of the aeroplane scouts and patrols. Under these circumstances it is suggested that balloon-mines should be sent aloft and be concealed in the clouds. It would be impossible to detect the wires holding them captive, so that the precise location of the lurking danger would not be divined by the invader. Of course, the chances are that the invading airship would unconsciously miss the mines; on the other hand the possibilities are equally great that it would blunder into one of these traps and be blown to atoms.

An English airman has recently suggested a means of mining invading Zeppelins which differs completely from the foregoing proposals. His idea is that aeroplanes should be equipped with small mines of the contact type, charged with high explosives, and that the latter should be lowered from the aeroplane and be trawled through the atmosphere. As an illustration I will suppose that a hostile aircraft is sighted by a patrolling aeroplane. The pilot's companion in the latter immediately prepares his aerial mine, fixing the detonator, and attaching the mine to the wire. The latter is then dropped overboard, the wire being paid out from a winch until it has descended to the level of the hostile craft. The airman now manoeuvres in the air circling about the airship, dragging his mine behind him, and endeavouring to throw it across or to bring it into contact with the airship below. Naturally the latter, directly it observed the airman's object, would endeavour to elude the pursuing trawling mine, either by crowding on speed or by rising to a greater altitude. The aeroplane, however, would have the advantage both in point of speed and powers of climbing, while there is no doubt that the sight of the mine swinging in the air would exert a decisive moral effect upon those in the airship.

Attempts to render the mine harmless by discharging it prematurely with the aid of rifle and machine-gun fire would, of course, be made by the crew of the airship, but the trawling mine would prove a very difficult target to strike. If such a missile were used against an airship of the proportions of a Zeppelin the mine would inevitably be trawled across the vessel sooner or later. Once the airship had been fouled, the aviator would merely have to drive ahead, dragging the wire and its charge across the gas-bag until at last one of the contact levers of the mine was moved by being dragged against some part of the vessel, when the mine would be exploded. In such operations the aviator would run a certain risk, as he would be more or less above the airship, and to a certain degree within the zone of the ultimate explosion. But there is no doubt that he would succeed in his "fishing" exploit within a very short time.

This ingenious scheme has already been tested upon a small scale and has been found effective, the trawling bomb being drawn across its target and fired by contact within a few minutes. The experiment seems to prove that it would be simpler and more effectual to attack a hostile aircraft such as a Zeppelin in this manner than to drop free bombs at random. Moreover, we cannot doubt that the sight of a mine containing even ten or twelve pounds of high explosive dangling at the end of a wire would precipitate a retreat on the part of an airship more speedily than any other combative expedient.

The advocate of this mine-trawling method, who is a well-known aviator, anticipates no difficulty in manoeuvring a mine weighing 30 pounds at the end of 300 feet of fine wire. Success depends in a great measure on the skill of the aviator in maintaining a constant tension upon the line until it falls across its objective.

The process calls for a certain manifestation of skill in manoeuvring the aeroplane in relation to the airship, judgment of distance, and ability to operate the aeroplane speedily. The rapid ascensional capability of the airship, as compared with that of the aeroplane, is a disadvantage, but on the other hand, the superior mobility and speed of the aeroplane would tell decisively for success.

Among the many wonders which the Krupp organisation is stated to have perfected, and which it is claimed will create considerable surprise, is the aerial torpedo. Many of the Krupp claims are wildly chimerical, as events have already proved, but there is no doubt that considerable effort has been expended upon this latest missile, for which the firm is said to have paid the inventor upwards of L25,000--$125,000. Curiously enough the projectile was perfected within gunshot of the British aerodrome of Hendon and is stated to have been offered to the British Government at the time, and to have met with a chilling reception. One fact, however, is well established. The inventor went to Germany, and submitted his idea to Krupp, by whom it was tested without delay. Upon the completion of the purchase, the great armament manufacturers did not fail to publish broadcast the fact that they had acquired a mysterious new terror of the skies. That was some three years ago, and in the interval the cleverest brains of the German firm have been steadily devoting their time and energies to the improvement of the missile, the first appearance of which was recorded, in a somewhat hazy manner, in the closing days of December.

While the exact mechanism of this missile is a secret, the governing principles of its design and operation are known to a select few technicians in this country. Strange to say, the projectile was designed in the first instance in the interests of peace and humanty, but while engaged upon his experiments the inventor suddenly concluded that it would be a more profitable asset if devoted to the grim game of war. At the time the military significance of the airship and the aeroplane were becoming apparent; hence the sudden diversion of the idea into a destructive channel.

This aerial torpedo is a small missile carrying a charge of high explosive, such as trinitrotoluene, and depends for its detonation upon impact or a time fuse. It is launched into the air from a cradle in the manner of the ordinary torpedo, but the initial velocity is low. The torpedo is fitted with its own motive power, which comes automatically into action as the missile climbs into the air. This self-contained energy is so devised that the maximum power is attained before the missile has lost the velocity imparted in the first instance, the result being that it is able to continue its flight in a horizontal direction from the moment it attains the highest point in its trajectory, which is naturally varied according to requirements. But there is no secret about the means of propulsion. The body is charged with a slow-burning combustible, in the manner of the ordinary rocket, whereby it is given a rapid rotary motion.

Furthermore it is stated to be fitted with a small gyroscope in the manner of the torpedo used in the seas, for the purpose of maintaining direction during flight, but upon this point there is considerable divergence of opinion among technicians, the general idea being that the torpedo depends upon an application of the principle of the ordinary rocket rather than upon a small engine such as is fitted to the ordinary torpedo. The employment of a slow combustible ensures the maintenance of the missile in the air for a period exceeding that of the ordinary shell. It is claimed by the Germans that this projectile will keep aloft for half-an-hour or more, but this is a phantasy. Its maintenance of flight is merely a matter of minutes.

The belated appearance of this much-lauded projectile and its restricted use suggest that it is unreliable, and perhaps no more effective than the aerial torpedo which appeared in the United States during the Spanish-American War, and proved a complete failure. An effective and reliable means of combating or frustrating a dirigible attack, other than by gun-fire or resort to the drastic remedy of ramming the enemy, has yet to be devised.

CHAPTER XVII

WIRELESS IN AVIATION

In a previous chapter the various methods of signalling between the ground and the airman aloft have been described. Seeing that wireless telegraphy has made such enormous strides and has advanced to such a degree of perfection, one naturally would conclude that it constitutes an ideal system of communication under such conditions in military operations.

But this is not the case. Wireless is utilised only to a very limited extent. This is due to two causes. The one is of a technical, the other of a strategical character.

The uninitiated, bearing in mind the comparative ease with which wireless installations may be established at a relatively small expense, would not unreasonably think that no serious difficulties of a technical character could arise: at least none which would defy solution. But these difficulties exist in two or three different fields, each of which is peculiarly complex and demands individual treatment.

In the first place, there is the weight of the necessary installation. In the case of the dirigible this may be a secondary consideration, but with the aeroplane it is a matter of primary and vital importance. Again, under present conditions, the noise of the motor is apt to render the intelligent deciphering of messages while aloft a matter of extreme difficulty, especially as these are communicated in code. The engine noise might be effectively overcome by the use of a muffler such as, is used with automobiles, but then there is the further difficulty of vibration.

This problem is being attacked in an ingenious manner. It is proposed to substitute for audible signals visual interpretations, by the aid of an electric lamp, the fluctuations in which would correspond to the dots and dashes of the Morse code. Thus the airman would read his messages by sight instead of by sound.

This method, however, is quite in its infancy, and although attractive in theory and fascinating as a laboratory experiment or when conducted under experimental conditions, it has not proved reliable or effective in aeronautical operations. But at the same time it indicates a promising line of research and development.

Then there are the problems of weight and the aerial. So far as present knowledge goes, the most satisfactory form of aerial yet exploited is that known as the trailing wire. From 300 to 700 feet of wire are coiled upon a reel, and when aloft this wire is paid out so that it hangs below the aeroplane. As a matter of fact,when the machine is travelling at high speed it trails horizontally astern, but this is immaterial. One investigator, who strongly disapproves of the trailing aerial, has carried out experiments with a network of wires laid upon and attached to the surface of the aeroplane's wings. But the trailing wire is generally preferred, and certainly up to the present has proved more satisfactory.

The greatest obstacle, however, is the necessary apparatus. The average aeroplane designed for military duty is already loaded to the maximum. As a rule it carries the pilot and an observer, and invariably includes a light arm for defence against an aerial enemy, together with an adequate supply of ammunition, while unless short sharp flights are to be made, the fuel supply represents an appreciable load. Under these circumstances the item of weight is a vital consideration. It must be kept within a limit of 100 pounds, and the less the equipment weighs the more satisfactory it is likely to prove, other things being equal.

The two most successful systems yet exploited are the Dubilier and the Rouget. The former is an American invention, the latter is of French origin. Both have been tested by the British Military Aeronautical Department, and the French authorities have subjected the French system to rigorous trials. Both systems, within their limitations, have proved satisfactory.

The outstanding feature of the Dubilier system is the production of sine waves of musical frequency from continuous current, thus dispensing with the rotary converter. The operating principle is the obtaining of a series of unidirectional impulses by a condenser discharge, the pulsating currents following one another at regular intervals at a frequency of 500 impulses per second, which may be augmented up to 1,000 impulses per second. The complete weight of such an apparatus is 40 pounds; the electric generator, which is no larger than the motor used for driving the ordinary table ventilating fan, accounts for 16 pounds of this total. Under test at sea, upon the deck of a ship, a range of 250 miles has been obtained. The British Government carried out a series of experiments with this system, using a small plant weighing about 30 pounds, with which communication was maintained up to about 20 miles.

In the French system the Reuget transmitter is employed. The apparatus, including the dynamo, which is extremely small, weighs in all 70 pounds. A small alternator of 200 watts and 100 volts is coupled direct to the aeroplane motor, a new clutch coupler being employed for this purpose. By means of a small transformer the voltage is raised to 30,000 volts, at which the condenser is charged. In this instance the musical spark method is employed.

The whole of the high tension wiring is placed within a small space so as not to endanger the pilot, while the transformer is hermetically sealed in a box with paraffin. The aerial comprises a trailing wire 100 feet in length, which, however, can be wound in upon its reel within 15 seconds. This reeled antenna, moreover, is fitted with a safety device whereby the wire can be cut adrift in the event of an accident befalling the aeroplane and necessitating an abrupt descent. With this apparatus the French authorities have been able to maintain communication over a distance of 30 miles.

In maintaining ethereal communication with aeroplanes, however, a portable or mobile station upon the ground is requisite, and this station must be within the radius of the aerial transmitter, if messages are to be received from aloft with any degree of accuracy and reliability. Thus it will be recognised that the land station is as important as the aeroplane equipment, and demands similar consideration.

A wide variety of systems have been employed to meet these conditions. There is the travelling automobile station, in which the installation is mounted upon a motor-car. In this instance the whole equipment is carried upon a single vehicle, while the antenna is stowed upon the roof and can be raised or lowered within a few seconds. If motor traction is unavailable, then animal haulage may be employed, but in this instance the installation is divided between two vehicles, one carrying the transmitting and receiving apparatus and the generating plant, the other the fuel supplies and the aerial, together with spare parts.

The motive power is supplied by a small air cooled petrol or gasoline motor developing eight horse-power, and coupled direct to a 2-kilo watt alternator. At one end of the shaft of the latter the disk discharger is mounted, its function being to break up the train of waves into groups of waves, so as to impart a musical sound to the note produced in the receiver. A flexible cable transmits the electric current from the generator to the wagon containing the instruments. The aerial is built up of masts carried in sections.

The Germans employ a mobile apparatus which is very similar, but in this instance the mast is telescopic. When closed it occupies but little space. By turning the winch handle the mast is extended, and can be carried to any height up to a maximum of about 100 feet. The capacity of these mobile stations varies within wide limits, the range of the largest and most powerful installations being about 200 miles. The disadvantage of these systems, however, is that they are condemned to territories where the ground at the utmost is gently undulating, and where there are roads on which four-wheeled vehicles can travel.

For operation in hilly districts, where only trails are to be found, the Marconi Company, has perfected what may be described as "pack" and "knapsack" installations respectively. In the first named the whole of the installation is mounted upon the backs of four horses. The first carries the generator set, the second the transmitting instruments, the third the receiving equipment, and the fourth the detachable mast and stays.

The generator is carried upon the horse's saddle, and is fitted with a pair of legs on each side. On one side of the saddle is mounted a small highspeed explosion motor, while on the opposite side, in axial alignment with the motor, is a small dynamo. When it is desired to erect the installation the saddle carrying this set is removed from the horse's back and placed upon the ground, the legs acting as the support. A length of shaft is then slipped into sockets at the inner ends of the motor and dynamo shafts respectively, thus coupling them directly, while the current is transmitted through a short length of flexible cable to the instruments. The mast itself is made in lengths of about four feet, which are slipped together in the manner of the sections of a fishing rod, and erected, being supported by means of wire guys. In this manner an antenna from 40 to 50 feet in height may be obtained.

The feature of this set is its compactness, the equal division of the sections of the installation, and the celerity with which the station may be set up and dismantled in extremely mountainous country such as the Vosges, where it is even difficult for a pack-horse to climb to commanding or suitable positions, there is still another set which has been perfected by the Marconi Company.

This is the "knapsack" set, in which the whole of the installation, necessarily light, small, and compact, is divided among four men, and carried in the manner of knapsacks upon their backs. Although necessarily of limited radius, such an installation is adequate for communication within the restricted range of air-craft.

Greater difficulties have to be overcome in the mounting of a wireless installation upon a dirigible. When the Zeppelin was finally accepted by the German Government, the military authorities emphasised the great part which wireless telegraphy was destined to play in connection with such craft. But have these anticipations been fulfilled? By no means, as a little reflection will suffice to prove.

In the first place, a wireless outfit is about the most dangerous piece of equipment which could be carried by such a craft as the Zeppelin unless it is exceptionally well protected. As is well known the rigidity of this type of airship is dependent upon a large and complicated network of aluminium, which constitutes the frame. Such a huge mass of metal constitutes an excellent collector of electricity from the atmosphere; it becomes charged to the maximum with electricity.

In this manner a formidable contributory source of danger to the airship is formed. In fact, this was the reason why "Z-IV" vanished suddenly in smoke and flame upon falling foul of the branches of trees during its descent. At the time the Zeppelin was a highly charged electrical machine or battery as it were, insulated by the surrounding air. Directly the airship touched the trees a short circuit was established, and the resultant spark sufficed to fire the gas, which is continuously exuding from the gas bags.

After this accident minute calculations were made and it was ascertained that a potential difference of no less than 100,00 volts existed between the framework of the dirigible and the trees. This tension sufficed to produce a spark 4 inches in length. It is not surprising that the establishment of the electric equilibrium by contact with the trees, which produced such a spark should fire the hydrogen inflation charge. In fact the heat generated was so intense that the aluminium metallic framework was fused. The measurements which were made proved that the gas was consumed within 15 seconds and the envelope destroyed within 20 seconds.

As a result of this disaster endeavours were made to persuade Count Zeppelin to abandon the use of aluminium for the framework of his balloon but they were fruitless, a result no doubt due to the fact that the inventor of the airship of this name has but a superficial knowledge of the various sciences which bear upon aeronautics, and fully illustrates the truth of the old adage that "a little learning is a dangerous thing." Count Zeppelin continues to work upon his original lines, but the danger of his system of construction was not lost upon another German investigator, Professor Schiitte, who forthwith embarked upon the construction of another rigid system, similar to that of Zeppelin, at Lanz. In this vessel aluminium was completely abandoned in favour of a framework of ash and poplar.

The fact that the aluminium constituted a dangerous collector of electricity rendered the installation of wireless upon the Zeppelin not only perilous but difficult. Very serious disturbances of an electrical nature were set up, with the result that wireless communication between the travelling dirigible and the ground below was rendered extremely uncertain. In fact, it has never yet been possible to communicate over distances exceeding about 150 miles. Apart from this defect, the danger of operating the wireless is obvious, and it is generally believed in technical circles that the majority of the Zeppelin disasters from fire have been directly attributable to this, especially those disasters which have occurred when the vessel has suddenly exploded before coming into contact with terrestrial obstructions.

In the later vessels of this type the wireless installation is housed in a well insulated compartment. This insulation has been carried, to an extreme degree, which indicates that at last the authorities have recognised the serious menace that wireless offers to the safety of the craft, with the result that every protective device to avoid disaster from this cause has been freely adopted.

The fact that it is not possible to maintain cornmunication over a distance exceeding some 20 miles is a severe handicap to the progressive development of wireless telegraphy in this field. It is a totally inadequate radius when the operations of the present war are borne in mind. A round journey of 200, or even more miles is considered a mere jaunt; it is the long distance flight which counts, and which contributes to the value of an airman's observations. The general impression is that the fighting line or zone comprises merely two or three successive stretches of trenches and other defences, representing a belt five miles or so in width, but this is a fallacy. The fighting zone is at least 20 miles in width; that is to say, the occupied territory in which vital movements take place represents a distance of 20 miles from the foremost line of trenches to the extreme rear, and then comes the secondary zone, which may be a further 10 miles or more in depth. Consequently the airman must fly at least 30 miles in a bee-line to cover the transverse belt of the enemy's field of operations. Upon the German and Russian sides this zone is of far greater depth, ranging up to 50 miles or so in width. In these circumstances the difficulties of ethereal communication 'twixt air and earth may be realised under the present limitations of radius from which it is possible to transmit.

But there are reasons still more cogent to explain why wireless telegraphy has not been used upon a more extensive scale during the present campaign. Wireless communication is not secretive. In other words, its messages may be picked up by friend and foe alike with equal facility. True, the messages are sent in code, which may be unintelligible to the enemy. In this event the opponent endeavours to render the communications undecipherable to one and all by what is known as "jambing." That is to say, he sends out an aimless string of signals for the purpose of confusing senders and receivers, and this is continued without cessation and at a rapid rate. The result is that messages become blurred and undecipherable.

But there is another danger attending the use of wireless upon the battlefield. The fact that the stations are of limited range is well known to the opposing forces, and they are equally well aware of the fact that aerial craft cannot communicate over long distances. For instance, A sends his airmen aloft and conversation begins between the clouds and the ground. Presently the receivers of B begin to record faint signals. They fluctuate in intensity, but within a few seconds B gathers that an aeroplane is aloft and communicating with its base. By the aid of the field telephone B gets into touch with his whole string of wireless stations and orders a keen look-out and a listening ear to ascertain whether they have heard the same signals. Some report that the signals are quite distinct and growing louder, while others declare that the signals are growing fainter and intermittent. In this manner B is able to deduce in which direction the aeroplane is flying. Thus if those to the east report that signals are growing stronger, while the stations on the west state that they are diminishing, it is obvious that the aeroplane is flying west to east, and vice versa when the west hears more plainly at the expense of the east. If, however, both should report that signals are growing stronger, then it is obvious that the aircraft is advancing directly towards them.

It was this ability to deduce direction from the sound of the signals which led to the location of the Zeppelin which came down at Lun6ville some months previous to the war, and which threatened to develop into a diplomatic incident of serious importance. The French wireless stations running south-east to north-west were vigilant, and the outer station on the north-west side picked up the Zeppelin's conversation. It maintained a discreet silence, but communicated by telephone to its colleagues behind.

Presently No. 2 station came within range, followed by Nos. 3, 4, 5, 6, and so on in turn. Thus the track of the Zeppelin was dogged silently through the air by its wireless conversation as easily and as positively as if its flight had been followed by the naked eye. The Zeppelin travellers were quite ignorant of this action upon the part of the French and were surprised when they were rounded-up to learn that they had been tracked so ruthlessly. Every message which the wireless of the Zeppelin had transmitted had been received and filed by the French.

Under these circumstances it is doubtful whether wireless telegraphy between aircraft and the forces beneath will be adopted extensively during the present campaign. Of course, should some radical improvement be perfected, whereby communication may be rendered absolutely secretive, while no intimation is conveyed to the enemy that ethereal conversation is in progress, then the whole situation will be changed, and there may be remarkable developments.

CHAPTER XVIII

AIRCRAFT AND NAVAL OPERATIONS

When once the flying machine had indicated its possibilities in connection with land operations it was only natural that endeavours should be made to adapt it to the more rigorous requirements of the naval service. But the conditions are so vastly dissimilar that only a meagre measure of success has been recorded. Bomb-throwing from aloft upon the decks of battleships appeals vividly to the popular imagination, and the widespread destruction which may be caused by dropping such an agent down the funnel of a vessel into the boiler-room is a favourite theme among writers of fiction and artists. But hitting such an objective while it is tearing at high speed through the water, from a height of several thousand feet is a vastly different task from throwing sticks and balls at an Aunt Sally on terra firma: the target is so small and elusive.

Practically it is impossible to employ the flying machine, whether it be a dirigible or an aeroplane, in this field. Many factors militate against such an application. In the first place there is a very wide difference between dry land and a stretch of water as an area over which to manoeuvre. So far as the land is concerned descent is practicable at any time and almost anywhere. But an attempt to descend upon the open sea even when the latter is as calm as the proverbial mill-pond is fraught with considerable danger. The air-currents immediately above the water differ radically from those prevailing above the surface of the land. Solar radiation also plays a very vital part. In fact the dirigible dare not venture to make such a landing even if it be provided with floats. The chances are a thousand to one that the cars will become water-logged, rendering re-ascent a matter of extreme difficulty, if not absolutely impossible. On the other hand, the aeroplane when equipped with floats, is able to alight upon the water, and to rest thereon for a time. It may even take in a new supply of fuel if the elements be propitious, and may be able to re-ascend, but the occasions are rare when such operations can be carried out successfully.

In operations over water the airman is confronted with one serious danger-- the risk of losing his bearings and his way. For instance, many attempts have been made to cross the North Sea by aeroplane, but only one has proved successful so far. The intrepid aviator did succeed in passing from the shore of Britain to the coast of Scandinavia. Many people suppose that because an airman is equipped with a compass he must be able to find his way, but this is a fallacy. The aviator is in the same plight as a mariner who is compelled from circumstances to rely upon his compass alone, and who is debarred by inclement weather from deciding his precise position by taking the sun. A ship ploughing the waters has to contend against the action of cross currents, the speed of which varies considerably, as well as adverse winds. Unless absolute correction for these influences can be made the ship will wander considerably from its course. The airman is placed in a worse position. He has no means of determining the direction and velocity of the currents prevailing in the atmosphere, and his compass cannot give him any help in this connection, because it merely indicates direction.

Unless the airman has some means of determining his position, such as landmarks, he fails to realise the fact that he is drifting, or, even if he becomes aware of this fact, it is by no means a simple straightforward matter for him to make adequate allowance for the factor. Side-drift is the aviator's greatest enemy. It cannot be determined with any degree of accuracy. If the compass were an infallible guide the airman would be able to complete a given journey in dense fog just as easily as in clear weather. It is the action of the cross currents and the unconscious drift which render movement in the air during fog as impracticable with safety as manoeuvring through the water under similar conditions. More than one bold and skilful aviator has essayed the crossing of the English Channel and, being overtaken by fog, has failed to make the opposite coast. His compass has given him the proper direction, but the side-drift has proved his undoing, with the result that he has missed his objective.

The fickle character of the winds over the water, especially over such expanses as the North Sea, constitutes another and seriously adverse factor. Storms, squalls, gales, and, in winter, blizzards, spring up with magical suddenness, and are so severe that no aircraft could hope to live in them. But such visitations are more to be dreaded by the lighter-than-air than by the heavier-than-air machines. The former offers a considerable area of resistance to the tempest and is caught up by the whirlwind before the pilot fully grasps the significant chance of the natural phenomenon. Once a dirigible is swept out of the hands of its pilot its doom is sealed.

On the other hand, the speed attainable by the aeroplane constitutes its safety. It can run before the wind, and meantime can climb steadily and rapidly to a higher altitude, until at last it enters a contrary wind or even a tolerably quiescent atmosphere. Even if it encounters the tempest head on there is no immediate danger if the aviator keep cool. This fact has been established times out of number and the airman has been sufficiently skilful and quick-witted to succeed in frustrating the destructive tactics of his natural enemy.

Only a short while ago in France, British airmen who went aloft in a gale found the latter too strong for them. Although the machine was driven full speed ahead it was forced backwards at the rate of 10 miles per hour because the independent speed of the aeroplane was less than the velocity of the wind. But a dirigible has never succeeded in weathering a gale; its bulk, area, and weight, combined with its relatively slow movement, are against it, with the result that it is hurled to destruction. All things considered, the dirigible is regarded as an impracticable acquisition to a fleet, except in the eyes of the Germans, who have been induced to place implicit reliance upon their monsters. The gullible Teuton public confidently believes that their Dreadnoughts of the air will complete the destruction of the British fleet, but responsible persons know full well that they will not play such a part, but must be reserved for scouting. Hitherto, in naval operations, mosquito water-craft, such as torpedo-boats, have been employed in this service. But these swift vessels suffer from one serious disability. The range of vision is necessarily limited, and a slight mist hanging over the water blinds them; the enemy may even pass within half-a-mile of them and escape detection.

The Zeppelin from its position 1,000 feet or more above the water, in clear weather, has a tremendous range of vision; the horizon is about 40 miles distant, as compared with approximately 8 miles in the case of the torpedo-boat. of course an object, such as a battleship, may be detected at a far greater range. Consequently the German naval programme is to send the Zeppelin a certain distance ahead of the battleship squadron. The dirigible from its coign of vantage would be able to sight a hostile squadron if it were within visual range and would communicate the fact to the commander of the fleet below. The latter would decide his course according to information received; thus he would be enabled to elude his enemy, or, if the tidings received from the aerial scout should be favourable, to dispose his vessels in the most favourable array for attack.

The German code of naval tactics does not foreshadow the use of dirigible aircraft as vessels of attack. Scouting is the primary and indeed the only useful duty of the dirigible, although it is quite possible that the aerial craft might participate in a subsequent naval engagement, as, indeed, has been the case. Its participation, however, would be governed entirely by climatic conditions. The fact that the dirigible is a weak unit of attack in naval operations is fully appreciated by all the belligerents.

The picture of a sky "black with Zeppelins" may appeal to the popular imagination, and may induce the uninitiated to cherish the belief that such an array would strike terror into the hearts of the foe, but the naval authorities are well aware that no material advantage would accrue from such a force. In the first place they would constitute an ideal target for the enemy's vessels. They would be compelled to draw within range in order to render their own attack effective, and promiscuous shooting from below would probably achieve the desired end. One or more of the hostile aircraft would be hit within a short while. Such disasters would undoubtedly throw the aerial fleet into confusion, and possibly might interfere with the tactical developments of its own friends upon the water below.

The shells hurled from the Zeppelins would probably inflict but little damage upon the warships beneath. Let it be conceded that they weigh about 500 pounds, which is two-thirds of the weight of the projectile hurled from the Krupp 128-centimetre howitzer. Such a missile would have but little destructive effect if dropped from a height of 1,000 feet. To achieve a result commensurate with that of the 28-centimetre howitzer the airship would have to launch the missile from a height of about 7,000 feet. To take aim from such an altitude is impossible, especially at a rapidly moving target such as a battle-cruiser.

The fact must not be forgotten that Count Zeppelin himself has expressed the opinion, the result of careful and prolonged experiments, that his craft is practically useless at a height exceeding 5,000 feet. Another point must not be overlooked. In a spirited naval engagement the combatants would speedily be obliterated from the view of those aloft by the thick pall of smoke--the combination of gun-fire and emission from the furnaces and a blind attack would be just as likely to damage friend as foe.

Even if the aircraft ventured to descend as low as 5,000 feet it would be faced with another adverse influence. The discharge of the heavy battleship guns would bring about such an agitation of the air above as to imperil the delicate equilibrium of an airship. Nor must one overlook the circumstance that in such an engagement the Zeppelins would become the prey of hostile aeroplanes. The latter, being swifter and nimbler, would harry the cumbersome and slow-moving dirigible in the manner of a dog baiting a bear to such a degree that the dirigible would be compelled to sheer off to secure ts own safety. Desperate bravery and grim determination may be magnificent physical attributes, ut they would have to be superhuman to face the stinging recurrent attacks of mosquito-aeroplanes.

The limitations of the Zeppelin, and in fact of all dirigible aircraft, were emphasised upon the occasion of the British aerial raid upon Cuxhaven. Two Zeppelins bravely put out to overwhelm the cruisers and torpedo boats which accompanied and supported the British sea-planes, but when confronted with well-placed firing from the guns of the vessels below they quickly decided that discretion was the better part of valour and drew off. In naval operations the aeroplane is a far more formidable foe, although here again there are many limitations. The first and most serious is the severely limited radius of action. The aeroplane motor is a hungry engine, while the fuel capacity of the tank is restricted. The German military authorities speedily realised the significance of this factor and its bearing upon useful operations, and forth with carried out elaborate endurance tests. In numerable flights were made with the express purpose of determining how long a machine could remain in the air upon a single fuel supply.

The results of these flights were collated and the achievements of each machine in this direction carefully analysed, a mean average drawn up, and then pigeon-holed. The results were kept secret, only the more sensational records being published to the world. As the policy of standardisation in the construction of aeroplanes was adopted the radius of action of each type became established. It is true that variations of this factor even among vessels exactly similar in every respect are inevitable, but it was possible to establish a reliable mean average for general guidance.

The archives of the Berlin military department are crowded with facts and figures relating to this particular essential, so that the radius of action, that is the mileage upon a single fuel charge, of any class and type of machine may be ascertained in a moment. The consequence is that the military authorities are able to decide the type of aeroplane which is best suited to a certain projected task. According to the dossier in the pigeon-hole, wherein the results of the type are filed, the aeroplane will be able to go so far, and upon arriving at that point will be able to accomplish so much work, and then be able to return home. Consequently it is dispatched upon the especial duty without any feeling of uncertainty.

Unfortunately, these experimental processes were too methodical to prove reliable. The endurance data were prepared from tests carried out in the aerodrome and from cross-country trials accomplished under ideal or fair-weather conditions. The result is that calculations have been often upset somewhat rudely by weather conditions of a totally unexpected character, which bring home vividly the striking difference between theory and practice.

The British and French aviation authorities have not adopted such methodical standardisation or rule of thumb inferences, but rather have fostered individual enterprise and initiative. This stimulation of research has been responsible for the creation of a type of aeroplane specially adapted to naval service, and generically known as the water plane, the outstanding point of difference from the aeroplane being the substitution of canoes or floats for the wheeled chassis peculiar to the land machine. The flier is sturdily built, while the floats are suf ficiently substantial to support the craft upon the water in calm weather. Perhaps it was the insular situation of the British nation which was responsible for this trend of development, because so far as Britain is concerned the sea-going aeroplane is in dispensable. But the salient fact remains that to-day the waterplane service of Great Britain is the most efficient in the world, the craft being speedy, designed and built to meet the rough weather conditions which are experienced around these islands, and ideal vessels for patrol and raiding duties.

So far as the British practice is concerned the waterplane is designed to operate in conjunction with, and not apart from, the Navy. It has been made the eyes of the Navy in the strictest interpretation of the term. In any such combination the great difficulty is the establishment of what may be termed a mobile base, inasmuch as the waterplane must move with the fleet. This end has been achieved by the evolution of a means of carrying a waterplane upon, and launching it from, a battleship, if necessary.

For this purpose a docking cradle or way has been provided aft where the aeroplane may be housed until the moment arrives for its employment.

Several vessels have been devoted to this nursing duty and are known as parent ships to the waterplane service. All that is requisite when the time arrives for the use of the seaplane is to lift it bodily by derrick or crane from its cradle and to lower it upon the water. It will be remembered that the American naval authorities made an experiment with a scheme for directly launching the warplane from the deck of a battleship in the orthodox, as well as offering it a spot upon which to alight upon returning from a flight, while Wing-Commander Samson, R.N., D.S.O., the famous British airman, repeated the experiment by flying from a similar launching way installed upon H.M.S. Hibernia. But this practice has many shortcomings. So far as the British and French navies are concerned, the former process is preferred. Again, when the waterplane returns from a flight it is admitted that it is simpler, quicker, and safer for it to settle upon the water near the parent ship and to be lifted on board.

As a sea-scout the waterplane is overwhelmingly superior to the dirigible as events have conclusively proved. Its greater mobility and speed stand it in excellent stead because it is able to cover a larger area within a shorter space of time than its huge and unwieldy contemporary. Furthermore, it is a difficult target to hit and accordingly is not so likely to be brought down by hostile fire. There is another point in its favour. The experience of the war has proved that the numerically inferior enemy prefers to carry out his naval operations under the cover of the mist and haze which settle upon the water, and yet are of sufficient depth to conceal his identity and composition. Such mists as a rule comprise a relatively thin bank of low-lying vapour, which while enveloping the surface of the water in an impenetrable pall, yet permits the mast-heads of the vessels to stand out clearly, although they cannot be detected from the water-level or even from the control and fighting tops of a warship. A scouting waterplane, however, is able to observe them and note their movement, and accordingly can collect useful information concerning the apparent composition of the hidden force, the course it is following, its travelling speed, and so forth, which it can convey immediately to its friends.

The aeroplane has established its value in another manner. Coal-burning vessels when moving at any pronounced speed invariably throw off large quantities of smoke, which may be detected easily from above, even when the vessels themselves are completely hidden in the mist. It was this circumstance which revealed the presence of the British squadron in the affair of the Bight of Heligoland.

The German airman on patrol duty from the adjacent base on the island of Heligoland detected the presence of this smoke, above the low-lying bank of fog, although there were no other visible signs of any vessels. Fully cognisant of the fact that the German Fleet was at anchor in a safe place he naturally divined that the smoke proceeded from a hostile squadron, evidently bent upon a raid. He returned to his headquarters, conveyed the intelligence he had collected to his superior officers, upon receipt of which a German cruiser squadron was sent out and engaged the British vessels to its own discomfiture. But for the airman's vigilance and smartness there is no doubt that the British squadron would have accomplished a great coup.

This incident, however, served to reveal that the aerial scout is prone to suffer from over-keenness and to collect only a partial amount of information. Upon this occasion the German watchman detected the presence of the British torpedo-boat and light cruiser force. Had he continued his investigations and made a wider sweep he would have discovered the proximity of the British battle-cruiser squadron which routed the German force, the latter having acted on incomplete information.

While the low-lying sea-fog is the navigator's worst enemy, it is the airman's greatest friend and protection. It not only preserves him against visual discovery from below, but is an excellent insulator of sound, so that his whereabouts is not betrayed by the noise of his motor. It is of in calculable value in another way. When a fog prevails the sea is generally as smooth as the pro verbial mirror, enabling the waterplanes to be brought up under cover to a suitable point from which they may be dispatched. Upon their release by climbing to a height of a few hundred feet the airmen are able to reach a clear atmosphere, where by means of the compass it is possible to advance in approximately the desired direction, safe from discovery from below owing to the fog. If they are "spotted" they can dive into its friendly depths, complete their work, and make for the parent ship.

Low-lying sea-fogs are favourable to aerial raids provided the scout is able to catch sight of the upper parts of landmarks to enable him to be sure of the correctness of his line of flight-in cases where the distance is very short compass direction is sufficiently reliable-because the bank of vapour not only constitutes a perfect screen, but serves as a blanket to the motor exhaust, if not completely, at least sufficiently to mislead those below. Fogs, as every mariner will testify, play strange tricks with the transmission of sound. Hence, although those on the vessels below might detect a slight hum, it might possibly be so faint as to convey the impression that the aviator was miles away, when, as a matter of fact, he was directly overhead. This confusion arising from sound aberration is a useful protection in itself, as it tends to lure a naval force lying in or moving through the fog into a false sense of security.

The development of the submarine revealed the incontrovertible fact that this arm would play a prominent part in future operations upon the water: a presage which has been adequately fulfilled during the present conflict. The instinct of self-preservation at once provoked a discussion of the most effective ways and means of disguising its whereabouts when it travels submerged. To this end the German naval authorities conducted a series of elaborate and interesting experiments off the island of Heligoland. As is well known, when one is directly above a stretch of shallow water, the bottom of the latter can be seen quite distinctly. Consequentiy, it was decided to employ aerial craft as detectives. Both the aeroplane and the dirigible took part in these experiments, being flown at varying heights, while the submarine was maneouvred at different depths immediately below. The sum of these investigations proved conclusively that a submarine may be detected from aloft when moving at a depth of from 30 to 40 feet. The outline of the submerged craft is certainly somewhat blurred, but nevertheless it is sufficiently distinct to enable its identity to be determined really against the background or bottom of the sea. To combat this detection from an aerial position it will be necessary inter alia to evolve a more harmonious or protective colour-scheme for the submarine. Their investigations were responsible for the inauguration of the elaborate German aerial patrol of harbours, the base for such aerial operations being established upon the island of Heligoland.

So far the stern test of war as applied to the science of aeronautics has emphasised the fact that as a naval unit the dirigible is a complete failure. Whether experience will bring about a modification of these views time alone will show, but it is certain that existing principles of design will have to undergo a radical revision to achieve any notable results. The aeroplane alone has proved successful in this domain, and it is upon this type of aerial craft that dependence will have to be placed.

CHAPTER XIX

THE NAVIES of THE AIR

Less than three years ago the momentous and spectacular race among the Powers of Europe for the supremacy of the air began. At first the struggle was confined to two rivals--France and Germany--but as time progressed and the importance of aerial fleets was recognised, other nations, notably Great Britain, entered the field.

Germany obtained an advantage. Experiment and research were taken up at a point which had been reached by French effort; further experiments and researches were carried out in German circles with secret and feverish haste, with the result that within a short time a pronounced degree of efficiency according to German ideals had been attained. The degree of perfection achieved was not regarded with mere academic interest; it marked the parting of the ways: the point where scientific endeavour com manded practical appreciation by turning the success of the laboratory and aerodrome into the channel of commercial manufacture. In other words, systematic and wholesale production was undertaken upon an extensive scale. The component parts were standardised and arrangements were completed with various establishments possessed of the most suitable machinery to perfect a programme for turning out aeronautical requirements in a steady, continuous stream from the moment the crisis developed.

The wisdom of completing these arrangements in anticipation is now apparent. Upon the outbreak of hostilities many German establishments devoted to the production of articles required in the infinite ramifications of commerce found themselves deprived of their markets, but there was no risk that their large plants would be brought to a standstill: the Government ordered the manufacture of aeroplane parts and motors upon an extensive scale. In this manner not only were the industrial establishments kept going, but their production of aeronautical requirements relieved those organisations devoted to the manufacture of armaments, so that the whole resources and facilities of these could be concentrated upon the supply of munitions of war.

In France the air-fleet, although extensive upon the outbreak of war, was somewhat heterogeneous. Experiment was still being pursued: no type had met with definite official recognition, the result being that no arrangements had been completed for the production of one or more standard types upon an elaborate scale comparable with that maintained by Germany. In fact some six months after the outbreak of war there was an appreciable lack of precision on this point in French military. Many of the types which had established their success were forbidden by military decree as mentioned in a previous chapter, while manufacturing arrangements were still somewhat chaotic.

Great Britain was still more backward in the new movement. But this state of affairs was in a measure due to the division of the Fourth Arm among the two services. A well-organised Government manufactory for the production of aeroplanes and other aircraft necessities had been established, while the private manufacturers had completed preparations for wholesale production. But it was not until the Admiralty accepted responsibility for the aerial service that work was essayed in grim earnest.

The allocation of the aerial responsibilities of Great Britain to the Admiralty was a wise move. Experience has revealed the advantages accruing from the perfection of homogeneous squadrons upon the water, that is to say groups of ships which are virtually sister-craft of identical speed, armament, and so on, thus enabling the whole to act together as a complete effective unit. As this plan had proved so successful upon the water, the Admiralty decided to apply it to the fleet designed for service in the air above.

At the time this plan of campaign was definitely settled Great Britain as an aerial power was a long way behind her most fomidable rival, but strenuous efforts were made to reduce the handicap, and within a short while the greater part of this leeway had been made up. Upon the outbreak of war Great Britain undoubtedly was inferior to Germany in point of numbers of aircraft, but the latter Power was completely outclassed in efficiency, and from the point of view of PERSONNEL. The British had developed the waterplane as an essential auxiliary to naval operations, and here was in advance of her rival, who had practically neglected this line of eeperiment and evolution, resting secure in the assurance of her advisers that the huge dirigibles would be adequate for all exigencies on the water.

Indeed, when war was declared, all the Powers were found more or less wanting so far as their aerial fleets were concerned. If Germany's huge aerial navy had been in readiness for instant service when she invaded Belgium, she would have overcome that little country's resistance in a far shorter time and with much less waste of life. It was the Belgians who first brought home to the belligerents the prominent part that aircraft were destined to play in war, and the military possibilities of the aeroplane. True, the Belgians had a very small aerial navy, but it was put to work without delay and accomplished magnificent results, ascertaining the German positions and dispositions with unerring accuracy and incredible ease, and thus enabling the commander of the Belgian Army to dispose his relatively tiny force to the best advantage, and to offer the most effective resistance.

Great Britain's aerial navy, while likewise some what small, was also ready for instant service. The British Expeditionary force was supported by a very efficient aerial fleet, the majority of the vessels forming which flew across the Channel at high speed to the British headquarters in France so as to be available directly military preparations were begun, and the value of this support proved to be inestimable, since it speedily demoralised the numerically superior enemy.

France, like Germany, was somewhat dilatory, but this was attributable rather to the time occupied in the mobilisation of the Fourth Arm than to lack of energy. There were a round 1,500 aeroplanes ostensibly ready for service, in addition to some 26 dirigibles. But the fleet was somewhat scattered, while many of the craft were not immediately available, being in the shops or in dock for repairs and overhaul. During the period of mobilisation the so-called standing military force was augmented by about 500 machines which were acquired from private owners. The aeroplane factories were also, overhauled and re-organised so as to be in a position to remedy the inevitable wastage, but these organisation efforts were somewhat handicapped by the shortage of labour arising from the call to arms. France, moreover, imperilled her aerial strength by forbidding the use of 558 machines which were ready for service.

Germany's aerial fleet was of similar proportions to that of her Gallic neighbour, but curiously enough, and in strange contrast, there appeared to be a lack of readiness in this ramification of the Teuton war machine. The military establishment possessed about 1,000 machines--active and reserve-- of which it is estimated 700 were available for instant service. During the period of mobilisation a further 450 machines were added to the fleet, drawn for the most part from private owners. So far as the dirigibles were concerned 14 Zeppelins were ready for duty, while others were under construction or undergoing overhaul and repair. A few other types were also in commission or acquired during mobilisation, bringing the dirigible force to 40 machines all told.

But the greatest surprise was probably offered by Russia. Very little was known concerning Russian activities in this particular field, although it was stated that large orders for machines had been placed with various foreign manufactories. Certain factories also had been established within the Empire, although the character of their work and its results and achievements were concealed from prying eyes. In Russia, however, an appreciable number of private aeroplanes were in operation, and these, of course, were placed at the disposal of the authorities the moment the crisis developed.

The British and French aeroplane manufacturers had been busy upon Russian orders for many months previous to the outbreak of hostilities, while heavy shipments of component parts had been made, the assembling and completion of the machines being carried out in the country. It is generally believed that upon the outbreak of war Russia had a fleet of 800 aeroplanes in hand, of which total 150 were contributed from private sources. Even the dirigible had not been overlooked, there being nearly 20 of these craft attached to the Russian Army, although for the most part they are small vessels.

In comparison with the foregoing large aerial navies, that of Great Britain appeared to be puny. At the moment Great Britain possesses about 500 machines, of which about 200 are waterplanes. In addition, according to the Secretary of the Admiralty, 15 dirigibles should be in service. Private enterprise is supported by the Government, which maintains a factory for the manufacture of these craft.

During the two years preceding the outbreak of war the various Powers grew remarkably reticent concerning the composition and enlargement of their respective aerial fleets. No official figures were published. But at the same time it is a well-known fact that during the year 1913 France augmented her flying force by no fewer than 544 aeroplanes. Germany was no less energetic, the military acquisition in this branch, and during the self-same year, approaching 700 machines according to the semi-official reports published in that country.

The arrangements concluded for the manufacture of additional craft during the war are equally remarkable. The principal factory in Germany, (now devoting its energies to the production of these craft, although in happier days its normal complement of 4,000 men were responsible for the production of another commercial article) possesses facilities for turning out 30 complete aeroplanes per week, according to the statement of its managing director. But it is averred that this statement is purposely misleading, inasmuch as during the first fortnight of the campaign it was producing over 50 aeroplanes per week. It must be remembered that Germany is responsible for the supply of the majority of such craft for the Austnan armies, that country purchasing these vessels in large numbers, because in the early days of the conflict it was notoriously weak in this arm. Since the declaration of war strenuous efforts have been made to remedy this state of affairs, particularly upon the unexpected revelation of Russia's aerial strength.

It is computed that upon the outbreak of war the various Powers were in the position to show an aggregate of 4,980 aircraft of all descriptions, both for active service and reserve. This is a colossal fleet, but it serves to convey in a graphic manner the importance attached to the adrial vessel by the respective belligerents. So far as Germany is concerned she is sorely in need of additional machines. Her fleet of the air has lost its formidable character, owing to the fact that it has to be divided between two frontiers, while she has been further weakened by the enormous lengths of the two battle-fronts.

Russia has been able to concentrate her aerial force, which has proved of incalculable value to the Grand Duke Nicholas, who has expressed his appreciation of the services rendered by his fliers. The French likewise have been favoured by Fortune in this respect. Their aerial navy is likewise concentrated upon a single frontier, although a pronounced proportion has been reserved for service upon the Mediterranean sea-board for co-operation with the fleet. France suffers, however, to a certain degree from the length of her battle-line, which is over 200 miles in length. The French aerial fleet has been particularly active in the Vosges and the Argonne, where the difficult, mountainous, and densely wooded country has rendered other systems of observation of the enemy's movements a matter of extreme difficulty. The Germans have laboured under a similar handicap in this territory, and have likewise been compelled to centre a considerable proportion of their aerial fleet upon this corner of the extended battlefield.

It is in this region that the greatest wastage has been manifest. I have been informed by one correspondent who is fighting in this sternly contested area, that at one time a daily loss of ten German machines was a fair average, while highwater mark was reached, so far as his own observations and ability to glean information were concerned by the loss of 19 machines during a single day. The French wastage, while not so heavy upon the average, has been considerable at times.

The term wastage is somewhat misleading, if not erroneous. It does not necessarily imply the total loss of a machine, such as its descent upon hostile territory, but includes damage to machines, no matter how slight, landing within their own lines. In the difficult country of the Vosges many aeroplanes have come to earth somewhat heavily, and have suffered such damage as to render them inoperative, compelling their removal from the effective list until they have undergone complete overhaul or reconstruction. Upon occasions this wastage has been so pronounced that the French aviators, including some of the foremost fliers serving with the forces, have been without a machine and have been compelled to wait their turn.

I am informed that one day four machines, returning from a reconnaissance in force, crashed successively to the ground, and each had to be hauled away to the repair sheds, necessitating withdrawal from service for several days. Unfortunately the French, owing to their decision to rule out certain machines as unsuited to military service, have not yet perfected their organisation for making good this wastage, although latterly it has been apprecably reduced by greater care among the aviators in handling their vessels.

The fast vessels of the French aerial fleet have proved exceptionally valuable.

With these craft speeds of 95 and 100 miles or more per hour have been attained under favourable conditions, and pace has proved distinctly advantageous, inasmuch as it gives the French aviators a superiority of about 40 per cent over the average German machine. It was the activity and daring of the French fliers upon these high speed machines which induced the German airmen to change their tactics. Individual effort and isolated raiding operations were abandoned in favour of what might be described as combined or squadron attack. Six or eight machines advancing together towards the French lines somewhat nonplussed these fleet French mosquito craft, and to a certain degree nullified their superiority in pace. Speed was discounted, for the simple reason that the enemy when so massed evinced a disposition to fight and to follow harassing tactics when one of the slowest French machines ventured into the air.

It is interesting to observe that aerial operations, now that they are being conducted upon what may be termed methodical lines as distinct from corsair movements, are following the broad fundamental principles of naval tactics. Homogeneous squadrons, that is, squadrons composed of vessels of similar type and armament, put out and follow roughly the "single line ahead" formation. Upon sighting the enemy there is the manoeuvring for position advantage which must accrue to the speedier protagonist. One then, witnesses what might almost be described as an application of the process of capping the line or "crossing the 'T.'" This tends to throw the slower squadron into confusion by bending it back upon itself, meanwhile exposing it to a demoralizing fire.

The analogy is not precisely correct but sufficiently so to indicate that aerial battles will be fought much upon the same lines, as engagements between vessels upon the water. If the manoeuvres accomplish nothing beyond breaking up and scattering the foe, the result is satisfactory in as much as in this event it is possible to exert a driving tendency and to force him back upon the lines of the superior force, when the scattered vessels may be brought within the zone of spirited fire from the ground.

Attacks in force are more likely to prove successful than individual raiding tactics, as recent events upon the battlefield of Europe have demonstrated more or less convincingly. An attack in force is likely to cause the defenders upon the ground beneath to lose their heads and to fire wildly and at random, with the result that the airmen may achieve their object with but little damage to themselves. This method of attacking in force was essayed for the first time by the British aerial fleet, which perhaps is not surprising, seeing that the machines are manned and the operations supervised by officers who have excelled in naval training, and who are skilled in such movements.

No doubt this practice, combined with the daring of the British aviators, contributed very materially to the utter demoralisation of the German aerial forces, and was responsible for that hesitancy to attack a position in the vicinity of the British craft which became so manifest in the course of a few weeks after the outbreak of hostilities.

One of the foremost military experts of the United States, who passed some time in the fighting zone, expressed his opinion that the British aerial force is the most efficient among the belligerents when considered as a unit, the French flier being described by the same authority as most effective when acting individually, owing to personal intrepidity. As a scout the French aviator is probably unequalled, because he is quick to perceive and to collect the data required, and when provided with a fast machine is remarkably nimble and venturesome in the air. The British aviators, however, work as a whole, and in the particular phases where such tactics are profitable have established incontestable superiority. At first the German aerial force appeared to possess no settled system of operation. Individual effort was pronounced, but it lacked method. The Germans have, however, profited from the lessons taught by their antagonists, and now are emulating their tactics, but owing to their imperfect training and knowledge the results they achieve appear to be negligible.

The dirigible still remains an unknown quantity in these activities, although strange to relate, in the early days of the war, the work accomplished by the British craft, despite their comparatively low speed and small dimensions, excelled in value that achieved by the warplanes. This was particularly noticeable in matters pertaining to reconnaissance, more especially at night, when the British vessels often remained for hours together in the air, manoeuvring over the hostile lines, and gathering invaluable information as to the disposition and movements of the opposing forces.

But it is probably in connection with naval operations that the British aerial fleet excels. The waterplanes have established their supremacy over the naval dirigible in a striking manner. British endeavour fostered the waterplane movement and has carried it to a high degree of perfection. The waterplane is not primarily designed to perform long flights, although such may be carried out if the exigencies demand. The practice of deputing certain vessels to art as "parent ships" to a covey of waterplanes has proved as successful in practice, as in theory. Again, the arrangements for conveying these machines by such means to a rendezvous, and there putting them into the water to complete a certain duty, have been triumphantly vindicated. At the time this idea was embraced it met with a certain degree of hostile criticism: it was argued that the association of the two fighting, machines would tend towards confusion, and impair the efficiency of both.

Practice has refuted this theory. The British aerial raids upon Cuxhaven and other places would have been impossible, and probably valueless as an effective move, but for the fact that it was possible to release the machines from a certain point upon the open sea, within easy reach of the cooperating naval squadron. True, the latter was exposed to hostile attack from submarines, but as results proved this was easy to repel. The aircraft were enabled to return to their base, as represented by the rendezvous, to be picked up, and to communicate the intelligence gained from their flight to the authorities in a shorter period of time than would have been possible under any other circumstances, while the risk to the airmen was proportionately reduced.

The fact that the belligerents have built up such huge aerial navies conclusively proves that the military value of the Fourth Arm has been fully appreciated. From the results so far achieved there is every indication that activity in this direction will be increased rather than diminished.

###

www.ingramcontent.com/pod-product-compliance
Lightning Source LLC
Chambersburg PA
CBHW070321190526
45169CB00005B/1691